CAMBRIDGE SOCIAL BIOLOGY TOPICS

Series editors
S. Tyrell Smith and Alan Cornwell

HEREDITY AND HUMAN DIVERSITY

Stephen Tomkins
Head of Biology, Hills Road Sixth Form College, Cambridge

CAMBRIDGE
UNIVERSITY PRESS

To O and U, with gratitude for more than genes.

Published by the Press Syndicate of the University of Cambridge
The Pitt Building, Trumpington Street, Cambridge CB2 1RP
40 West 20th Street, New York, NY 10011–4211, USA
10 Stamford Road, Oakleigh, Melbourne 3166, Australia

First published 1989
Third printing 1993

Printed in Great Britain at the University Press, Cambridge

British Library cataloguing in publication data
Tompkins, Stephen
  Heredity and human diversity
  1. Man. Genetics.
  I. Title
  573.2'1

ISBN 0 521 31229 9   SE

Author's acknowledgements
I should like to thank Michael Reiss and Alan Cornwell, in particular, for their help
with the manuscript of this book. I should also like to express my appreciation to
the following for their assistance, encouragement, patience, criticism or good
counsel: Myles Axton, Peter Bilton, Kay E. Davies, Clare Davison, Harriet Elson,
Katie Joyce, Spencer Hagard, Mike Hobart, Edward Holden, Katherine Pate, Ninette
Premdas, Lucy Purkis, Tony Seddon, David Summers, Helga Tomkins, Merlin and
Joseph Tomkins, Peter Thomas, Tom Wakeford, Mary Warnock and John
Whitmore.
The publishers would like to thank the following for supplying photographs:
Fig. 1.3 Professor U. Laemmli, University of Geneva, Switzerland; Fig. 2.1 Dr A.
McDermott, S.W. Regional Cytogenetics Service, Bristol; Fig. 2.3. Dr A. Mullinger
and Dr B. Johnson, Dept. of Zoology, University of Cambridge; Fig. 5.3 Muscular
Dystrophy Group of Great Britain and Northern Ireland; Fig. 5.7 BBC Hulton
Picture Library, London; Fig. 6.6 From *Journal of Heredity*, Oxford University Press
© 1914; Fig. 7.3. Lionel Willatt, East Anglian Regional Genetic Service,
Addenbrooke's Hospital, Cambridge; Fig. 8.7 Myles Axton, Imperial College,
London; Fig. 10.7 Janine Wiedel, London
Cover photograph reproduced by kind permission of the Down's Syndrome
Association, photographer: Bob Bray
The publishers would like to thank the following for permission to redraw diagrams:
Fig. 4.3. Kalmus, *Annals of Eugenics*, 15, 24–48, Cambridge University Press; Fig.
7.2. and Fig. 8.5. J.B. Jenkins, *Human Genetics*, copyright © 1983, Benjamin/
Cummings Publishing Co.; Fig. 10.5 *Nature*, vol. 190, p1179 copyright © 1961,
Macmillan Magazines Ltd.

Contents

Preface

This book aims to set out in a small space an introduction to human genetics. The dimensions of the subject are very wide, ranging from the molecular intricacies of DNA, at one extreme, to the global diversity of human beings at the other. Parts of the book are necessarily concentrated and most student readers will be helped by having a standard elementary biology text as a companion.

This book does not just aim to be an introductory text to human genetics, for as soon as one has learnt the basic facts about heredity one realises that knowledge of genetics has implications for the lives of people. This is very much the philosophy behind the subject of Social Biology, for it is the biology of the human species in all its dimensions. Human genetics is one of the most fascinating of its areas and is perhaps still the least understood.

Traditionally heredity has been taught from a foundation in Mendelian theory. However, this book starts with the story of DNA, the stuff of genes. The long coded molecule, first described by Crick and Watson in 1953, is now more visible, thanks to ultra-microscopy, and better understood, thanks to biochemical genetics, than just a few years ago. Our chromosomes, in which the DNA is stored are described and the way that their coupling behaviour and separation generates diversity, is explained. The sex cells and the embryos that come from them have a genetic individuality. From both our genes and developmental environment comes the uniqueness of each human being. Although the cell has faithful mechanisms for copying and repair of DNA our genes may be damaged and our chromosomes broken. Such mutation brings changes: both the new variations that improve a species, under natural selection, and also the harmful variants that are the cause of hereditary disease. The book concludes with a close look at such questions as race, genetic counselling, eugenics, euthanasia, the test-tube baby revolution and abortion. Young people need not only the facts to judge such things for themselves, but also a grounding in ethical debate.

This book is designed for sixth form students following A and AS level courses in Biology and Social Biology. It may also be found useful for students following para-medical and pre-medical courses. Additionally it is intended for those whose interest in contemporary social and moral issues requires a foundation in science.

1   The nucleic acids: coded information for life

1.1   Introduction

Deoxyribonucleic acid, **DNA**, is the chemical compound which carries the hereditary information. This genetic substance underpins the full diversity of animal and plant life. DNA carries its information in a coded form, with a meaning that it is often hard to interpret. Outside the living cell and organism DNA is a mere chemical, but inside it is transforming in its influence. The information that it carries can be regarded as a set of instructions or a blueprint for the development of a human life. Our DNA came to us from our two parents and they received theirs from their parents. From countless generations back in time our DNA reflects our ancient origin and evolution.

DNA is found in all living organisms and comprises about 5% of the dry weight of human cells. It has been nicknamed 'the thread of life' for the molecule may be several millimetres long although itself very very thin. DNA is wound up and packaged by coiling and folding into the tight structure of our chromosomes. It is so thin that if the nuclear DNA of a single cell were to be unwound fully it would be a whole metre in length; the total length of just one person's DNA would extend for many millions of miles. DNA is the stuff of genes and as such is the inescapable starting point for a book on human genetics.

Genetics is the science of heredity. As a scientific discipline it dates only from the start of the twentieth century when it was first realised that the particles of inheritance, the **genes**, are passed on unchanged from one generation to another. For a long time the nature of the gene, even in its chemical nature, was a mystery, but in the latter half of the twentieth century there has been no uncertainty that DNA is the genetic substance. Before we may see how genetic information is written into DNA molecular structure it is essential to be clear about the nature of code systems themselves.

1.2   Codes

When reading a text book one does not expect to find **hmbnlopdgdmrhakd bncdc vnqcr**. Such seeming nonsense is difficult to read and one might suppose that the type-setting machine used in the book's production had been misprogrammed. In fact the text has just been 'ciphered'. Secret service agents may use a cipher or coding device to put a written passage into such a seemingly meaningless form. The cipher used in this case merely involves shifting letters by one position in the alphabetical sequence; B becomes A, C becomes B, D becomes C and so on. In this particular case therefore–

HMBNLOPDGDMRHAKD   BNCDC   VNQCR

reads as

INCOMPREHENSIBLE   CODED   WORDS.

(Now re-read the first sentence of the paragraph at the bottom of the last page with the last three words deciphered.)

We may take the language coding analogy further, for writing is a coded form of our spoken language. Our speech is a coded language built up within our culture. Other languages are in different code. The word *mbwa* may mean nothing to you, but translated into *perro, hund* or *chien* it may be more familiar. The Swahili, Spanish, German and French languages, respectively, may convey nothing unless you can decode and transcribe the meaning into your own language.

Codes also require punctuation to be intelligible.

Compare

'The dog ate, the cat then slept soundly.'

with

'The dog ate the cat, then slept soundly.'

The forward or backward direction of reading matters too.

Compare

DOG and GOD

Our lives are so full of codifications of meaning that we take them for granted. Ours, we are told, is the age of 'information technology', for information is not only written but is recorded on discs, on magnetic tape and is digitised in circuits and pulses of light, or is transmitted on the air waves. It is perhaps no accident that this age has also seen the discovery of **coded genetic information**.

1.3   The meaning of the genetic code is protein

The DNA code is translated in the formation of **proteins** and **polypeptides** in cells. For a long time proteins have been recognised as being very complex and delicate biochemicals. They are only produced by living things. Proteins are not only used to make up the many structural components of cells but they are also used to make the vast variety and number of **enzymes** contained in the cell cytoplasm. These are attached to its membranes or found in its membrane bound spaces. The enzymes are the components of the cell which refashion other molecules, catalysing the reactions that change one substance into another. Enzymes also make possible the controlled release of energy in respiration, energy which itself is then directed to useful work. The systems of enzyme control, which constitute catabolism (breaking-down reactions) and anabolism (building-up reactions), are described together as **metabolism**. At a purely cellular and biochemical level life is a metabolic process. Enzymes make metabolic change possible. Because cells behave and interact according to their own chemistry, enzymes also indirectly control much of what an organism can achieve. Proteins are generally synthesised within the cells that contain them.

Thus DNA, which governs the nature of the enzymes synthesised, is the governor of both the cell and of the developmental programme of the multicellular individual.

One might be tempted to imbue DNA with some mystical power or status but it is not magical in itself, any more than is a copy of a Shakespeare play or an unplayed recording of a Beethoven symphony. As with the writings of Shakespeare and Beethoven, much DNA coding is unique, but like them it did not come together by one chance event alone. You may have only a metre length of DNA coding in each body cell coding for all of your metabolic functions, but it represents millions of years of innovation and evolutionary tuning to your ancestors' requirements. The lettered sequence of our DNA code is immense; it is longer than all the letters in 750 000 copies of this book. It is so long a code that you could not read out a half of it in your lifetime.

1.4 The discovery of DNA

Deoxyribonucleic acid was isolated by a Swiss biochemist, Freidrich Miescher, over a century ago. He was the first to show that nuclei possess a compound containing a five carbon sugar called **deoxyribose**, organic **nitrogenous bases** and acid **phosphate groups**. Ribose is a pentose sugar with the chemical formula $C_5H_{10}O_5$. Deoxyribose has one atom of oxygen less. Ribose also forms nucleic acids; these are different from DNA and are known as RNA.

Miescher had a clear concept of what the hereditary material might be like, but biochemistry at that early date was too crude a science to get him much further. In 1892 he wrote, in a letter of unwitting foresight, that some of the largest molecules encountered in living things, composed of similar but not identical pieces, could express all the rich variety of the hereditary message 'just as the words and concepts of all languages can find expression in the twenty to thirty letters of the alphabet'. Miescher clearly expected that this role would be fulfilled by proteins. It was half a century before his DNA discovery came to the fore.

In the early part of this century the **gene concept** became clear but the chemical nature of the gene was still a mystery. It was not until 1944 that Oswald Avery discovered that DNA, not protein, was the component in bacteria that was able to alter their genetic character. This discovery stimulated fresh research on DNA using some of the new techniques that had been found useful in the elucidation of protein structure. Erwin Chargaff, at Columbia University, analysed the nature and variety of the nitrogenous bases, whilst at King's College, London, Maurice Wilkins and Rosalind Franklin carried out careful X-ray crystallographic studies of the DNA molecule. It was in using the data gathered by these other scientists that James Watson and Francis Crick, at Cambridge University in 1953, suggested a double helical structure for DNA. (See the Bibliography, Watson, 1968.)

1.5 The double helix

DNA is a **polymer**, a large molecule of many repeated units. The units are called **nucleotides**. There are many other polymers in living things; proteins, for example are made up of amino acids. Single nucleotides have the three subunits that Miescher had recognised: an acid **phosphate group**, the **pentose** sugar deoxyribose, and one **nitrogenous base**. There are four bases in DNA, **adenine, guanine, cytosine** and **thymine**. The first two of these are larger units classed as **purines** whilst the latter two are slightly smaller units classed as **pyrimidines**. Cells contain many free nucleotides but most are polymerised together as nucleic acids. These polynucleotides form chains, each nucleotide forming strong covalent bonds between the sugar unit of one and the phosphate group of the next. A single strand of DNA is a string of nucleotides joined in a strong **sugar–phosphate chain** (see Fig. 1.1).

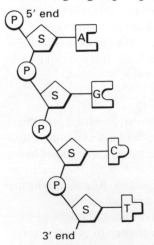

Fig. 1.1   A single sugar phosphate chain

Erwin Chargaff, in 1950, recognised that there were four types of base in DNA but that they occurred in varied proportions in different species. What was constant between all forms of life that he looked at was the correspondence between the amounts of adenine and thymine and the amounts of guanine and cytosine (see Table 1.1). Watson and Crick suggested that this

Table 1.1 The base equivalences of DNA, units are moles percent of bases

Source of DNA	Purines		Pyrimidines	
	adenine	guanine	cytosine	thymine
gut bacterium (Escherichia coli)	24.7	26.0	25.7	23.6
yeast	31.3	18.7	17.1	32.9
wheat	27.3	22.7	22.8	27.1
broad bean	29.7	20.6	20.1	29.6
salmon	29.7	20.8	20.4	29.1
bull	28.6	22.2	22.0	27.2
human	30.9	19.9	19.8	29.4

Source: Chargaff, E. (et al.) 1953, *Nature*, London, no. 172, p. 289

complementarity, or matching up, reflected a pairing reality in the molecule. If adenine (A) paired with thymine (T) then they would occur in equal amounts. Similarly if guanine (G) paired with cytosine (C) then they would occur in the same quantities also. Watson and Crick thus envisaged not one but two polynucleotide chains running parallel, united by their paired bases at the centre of a ladder-like form. (Because the molecular orientation of the sugars in the two molecular strands run in opposite directions, this alignment is described as 'anti-parallel'.) DNA is most simply regarded as being two long polynucleotide chains 'zipped' together. The X-ray crystallographic studies suggested a twisted helical form to the molecule, with ten base pairs stacked on the axis for each 360° revolution. The bonds between the chains that attach the bases together in pairs are weak hydrogen bonds, whilst those along the side chains are stronger covalent ones. The helical structure of the molecule lends it stability. Fig. 1.2 gives the dimensions and form of the molecule.

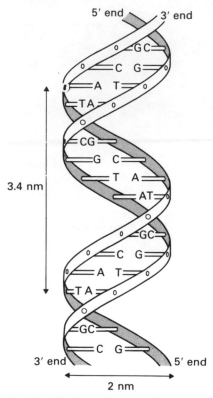

Fig. 1.2   The three dimensional structure of DNA

Fig. 1.3   Electron micrograph of DNA spilling out of a single disrupted human chromosome. The pattern here is unnatural because the chromosome proteins, including histones, have been largely removed. This picture shows about a quarter of the DNA in one chromosome.

The structure of DNA is not only elegant but also highly functional. The weak bondings between the bases allow an unzipping of the strand in a lengthwise direction to expose the coded language of the bases along its central axis; but as a helically coiled thread it is also a highly stable molecule that will store its coded information in a protected manner.

The **Watson and Crick model** was immediately seen to be the potentially code-carrying chemical that geneticists had long sought to find. The sugar–phosphate chain was strong and invariable, being the same in every molecule, but attached along each strand was a string of bases written in an alphabet of four letters A, G, C and T. The genius of Watson and Crick was to see that the sequence in which these bases occurred might be a language in code. This idea, of the code-carrying molecule, might only be plausible if the molecule could accomplish on a tiny scale what living cells can do on a vastly greater one, namely to divide into two with each half having an identical set of genetical information. If one considers for a moment that identical twins arise by a division of one zygote into two cells (and that each develops into two identical individuals), do we, in DNA, have a molecule that can replicate and carry on such a coded message of inheritance, both identically and without error, to each new cell? It seems that DNA can meet this challenge.

1.6 Semi-conservative replication by complementary base pairing

Watson and Crick noted that the larger bases, the purines, could not pair with each other. Adenine and guanine together would be too large for the calculated width of the molecule and the hydrogen bonds between them would not match up correctly. The smaller pyrimidines would be unable to bridge across the chain and did not match in their hydrogen bonding either. If a larger purine always paired with a smaller pyrimidine, the inter-chain distances would be constant and the helical structure regular and stable. Moreover, the hydrogen bond number and positions between adenine and thymine are such that only these two may bond together; adenine and cytosine just will not pair up. For similar reasons, it is found that guanine and cytosine form a stable pair, guanine being a mis-match with thymine. Bases thus each have their **complementary pair:** A with T and G with C.

It did not escape Watson and Crick's notice that the pairing of A with T and G with C provides a precise replication of the nucleotide sequence should each of the two separated strands subsequently rebuild their halves. Fig. 1.4 shows a sequence of just ten nucleotide pairs in a short length of molecule undergoing replication. The statistical probability of such a series of ten units forming *the same series* of bases by chance alone is, mathematically, the number of base choices at each addition (4), to the power of the number of choices made (10). As $4^{10} = 1\,048\,576$, for even such a short length there is only a one in a million probability of it coming together in this form by chance alone *unless* it is copied as Watson and Crick suggested.

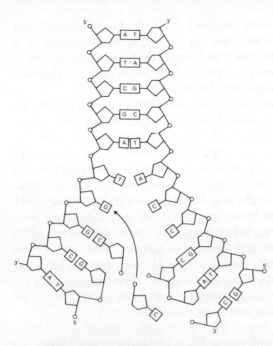

Fig. 1.4 A sequence of ten nucleotide pairs replicating. One strand is separated from the other in replication. Free nucleotides join into the old strand to build a new one, following the A–T, G–C pairing rule.

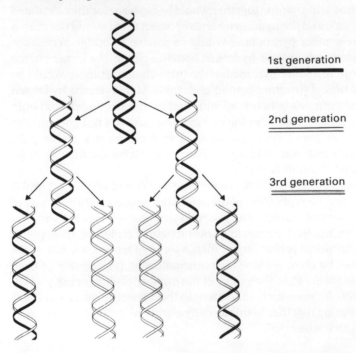

1st generation

2nd generation

3rd generation

Fig. 1.5 Meselson and Stahl's conclusion: in each generation of DNA a new strand is added to the old.

Within ten years of the first announcement of the structure of DNA, Meselson and Stahl had conducted an experiment that strongly supported this **semi-conservative replication hypothesis**. Their findings are summarised by Fig. 1.5. New DNA strands are formed on old ones by the condensation of free energised nucleotides into exactly complementary positions. This process is under the control of many enzymes and takes place between cell divisions. **DNA replication** takes place in all growing cells and is described fully in Chapter 2.

1.7 Deciphering the code

If the sequence of the nucleotides, with their four possible bases, represents a coded message in DNA, how is it to be read, deciphered and given meaning? Firstly, one should not underestimate the amount of genetic information that exists to be decoded from a cell. The metre length of DNA in a human cell contains three thousand million (3×10^9) base letters in a sequence. This amount of code may be expressed as three million **kilobases**, a kilobase being a thousand bases in sequence. (It is useful to think in kilobases because they represent lengths that are approximately equivalent to a small gene).

If the unique base sequence of each DNA molecule consists of coded messages, what is the cipher? How is the code given meaning? The cracking of the genetic code is principally attributed to Francis Crick, but would not have been achieved without the cooperative work of many scientists. In the 1960s the knowledge that proteins consisted of precise sequences of amino acids was recognised. Crick's challenge was to understand how DNA's four letter code could be read as signal instructions for the assembly of the twenty variable amino acid units of all proteins. At first sight there would not seem to be enough base varieties in DNA but Crick argued that the four letter code might be read in multiple groups of letters to give the precision necessary to exactly construct the amino acid sequence. If bases were to be read off in twos there would be insufficient combinations $(4 \times 4 = 16)$ to exactly specify each of twenty amino acids. Should bases be read off in threes there would be more than enough combinations $(4 \times 4 \times 4 = 64)$. It is now certain that the cipher reads in blocks of three as a **triplet code**, three bases in sequence standing for each specific amino acid. These triplets also code for the punctuation that stops each message sequence.

DNA on its own cannot give rise to proteins. It became very clear to these first molecular biologists that other cell machinery was involved, after all the nucleus is bounded by a membrane and protein synthesis takes place entirely outside the nucleus in the cytoplasm. Crick realised that there must be a means by which the coded message could be interpreted in the assembly of a protein. We now know that this is achieved by the action of a second molecule, related to DNA called **ribonucleic acid RNA**. RNA is the most important molecule in the decoding process for it transcribes, or literally writes out in a different form, the code into a sequence which can be translated into functional protein on the ribosomes of the cell cytoplasm. To understand the deciphering of the genetic code one must appreciate the various roles of RNA.

1.8 Ribonucleic acid

RNA is found in all cells. Like DNA it is a nucleotide polymer, but it has, as its name implies, ribose sugar instead of deoxyribose sugar in its structure. Its nucleotide polymer form is very varied but it is generally much shorter than DNA, single stranded and contains the different and more easily made pyrimidine **uracil** (U) as a substitute for thymine (T). RNA is less stable than the double helical DNA; its shorter life in the cell is one engaged in prodigious activity. RNA comes in three forms; **messenger RNA (mRNA)**, **transfer RNA (tRNA) and ribosomal RNA (rRNA)**. Their precise roles are inter-linked in protein synthesis in the two step processes of **transcription** and **translation**. These two processes are illustrated as they occur in the cell in Fig. 2.7 under section 2.8 in the next chapter.

1.9 DNA transcription

In the cells of higher organisms DNA is almost entirely confined to the nucleus. One sort of RNA acts as a messenger, copying the code from a length of DNA and taking it from the nucleus to a protein assembly site in the cytoplasm of the cell. This process of **transcription** into mRNA code has been most fully studied in bacteria. Here a single enzyme, **RNA polymerase**, is responsible. This enzyme binds to a sequence called 'the promoter' which signals exactly whereabouts on the chain copying should begin. The RNA polymerase enzymes start the transcription by unwinding the DNA and then building a complementary strand of RNA upon one of the two strands. The DNA strand that the enzyme binds with is dictated by the promoter and only one of the two strands is transcribed in a direction set by the orientation of the molecules in the chain. Note that the base pairing in RNA now substitutes uracil for thymine.

DNA code   C T A G T T A A G C A T A C C A C T
RNA code   G A U C A A U U C G U A U G G U G A

RNA polymerase continues to track along the DNA molecule transcribing the code into this new form. The newly formed RNA polynucleotide chain only binds weakly to the DNA strand and is soon released as a single strand of RNA. The RNA polymerase enzyme eventually arrives at a termination signal on the DNA strand at which moment it lets go of the DNA molecule. The latter then reforms its double helical structure. The whole process is rapid; a five kilobase gene may be transcribed in less than three minutes, at a rate of 30 bases a second. The single strand of messenger RNA now has the information in triplet code that is required to make a protein. (The way in which this final translation of code takes place is described at the end of Chapter 2).

By 1967 the genetic code had been deciphered fully and has since been found to be universal to life on Earth. The genetic code cipher is set out in Table 1.2.

The RNA triplets may be read out from their first, second and third letters; these are called **codons**. Note that several different codons may code for the same amino acid. The code must be read linearly and three punctuation stop signals are included. The 'central dogma' of molecular biology may be summed up in Francis Crick's classic phrase, '**DNA makes RNA makes protein**'.

Table 1.2 The genetic code: the triplet code is read in RNA base triplets (codons). A sequence of three nucleotide bases are translated into one amino acid in protein synthesis. The position of each letter in a triplet may be read from the table to give the abbreviated amino acid name.

1st position (5′ end) ↓	2nd position U	C	A	G	3rd position (3′ end) ↓
U	Phe	Ser	Tyr	Cys	U
	Phe	Ser	Tyr	Cys	C
	Leu	Ser	STOP	STOP	A
	Leu	Ser	STOP	Trp	G
C	Leu	Pro	His	Arg	U
	Leu	Pro	His	Arg	C
	Leu	Pro	Gln	Arg	A
	Leu	Pro	Gln	Arg	G
A	Ile	Thr	Asn	Ser	U
	Ile	Thr	Asn	Ser	C
	Ile	Thr	Lys	Arg	A
	Met	Thr	Lys	Arg	G
G	Val	Ala	Asp	Gly	U
	Val	Ala	Asp	Gly	C
	Val	Ala	Glu	Gly	A
	Val	Ala	Glu	Gly	G

A table of amino acid codes

Alanine (Ala) GCA GCC GCG GCU
Arginine (Arg) CGA CGC CGG CGU AGA AGG
Asparagine (Asn) AAU AAC
Aspartic Acid (Asp) GAU GAC
Cysteine (Cys) UGC UGU
Glutamic Acid (Glu) GAA GAG
Glutamine (Gln) CAA CAG
Glycine (Gly) GGA GGC GGG GGU
Histidine (His) CAC CAU
Isoleucine (Ile) AUU AUC AUA
Leucine (Leu) CUA CUC CUG CUU UUA UUG
Lysine (Lys) AAA AAG
Methionine (Met) AUG
Phenylalanine (Phe) UUU UUC
Proline (Pro) CCA CCC CCG CCU
Serine (Ser) AGC AGU UCA UCC UCG UCU
Threonine (Thr) ACA ACC ACG ACU
Tryptophan (Trp) UGG
Tyrosine (Tyr) UAC UAU
Valine (Val) GUA GUC GUG GUU
Stop UAA UAG UGA

Use the genetic code table to translate the transcribed RNA code on the opposite page. This short peptide has only five amino acids. Many proteins have more than a thousand.

2 The government of the cell

2.1 The nucleus rules the cell

Each human cell contains a set of genetic information that has been passed to it from its antecedent cells. Rudolph Virchow, in 1855, perceived this continuity of life by stating that 'all cells arise from pre-existing cells'. The immortality of the cell line is made plain for us by our own lives. Despite the fact that vast numbers of cells in our bodies die every day and that we ourselves, as collections of cells, must come to an end, most of us will pass on sex-cells to our children which will then multiply again in a new life. Virchow's idea is important and is not far removed from an idea in contemporary biology that DNA is selfishly ensuring its own immortality through the lives of its possessors. The 'selfish gene' hypothesis (Dawkins 1976) argues that even if one has no children oneself one's relatives, to whom one gives the greatest care, may well reproduce themselves and so one's own DNA will be passed on. What we hand down of course will not be the same molecules that we got from our parents, for those are bound to be lost, but they will be faithful replicas. It is this which allows the physical features in a family to be traced back to the parents or grandparents and beyond.

The first cell in an individual's life is the **zygote**, the fusion product of an egg and a single sperm. The maternal cytoplasm of that cell is already fully equipped with many organelles. Initially, at fertilisation, there are two nuclei; one is maternal and the other from the sperm head is paternal. Within a short while, at the first cell division, these two contributory nuclei divide side by side and the daughter nuclei go to opposite poles of the zygotic cell and fuse together to make single membrane bounded nuclei for the first time. From then on, throughout one's life, this new genetic nucleus, which is half maternal and half paternal in origin, has its DNA faithfully copied. The long threads of DNA are packaged together in the nucleus as numerous dark staining structures called **chromosomes**. The number of these hereditary units varies greatly between different species but for any one species the number is constant. Human body cells have exactly 46 per nucleus in each cell. Such a cell nucleus is described as **diploid** for it contains, within this number, two sets of chromosomes. In the human species each of the parents contribute 23 chromosomes. The diploid number is therefore 46. It is only the eggs and sperms, the germ cells, which have a single **haploid** set of 23. (It is common to express haploid and diploid numbers algebraically. The haploid number is n, the diploid number is $2n$. In humans, $n = 23$, and therefore $2n = 46$).

If a sample of living human cells is placed in a suitable culture medium for 2 to 3 days the cells may begin to divide. It is then possible to make microscopic

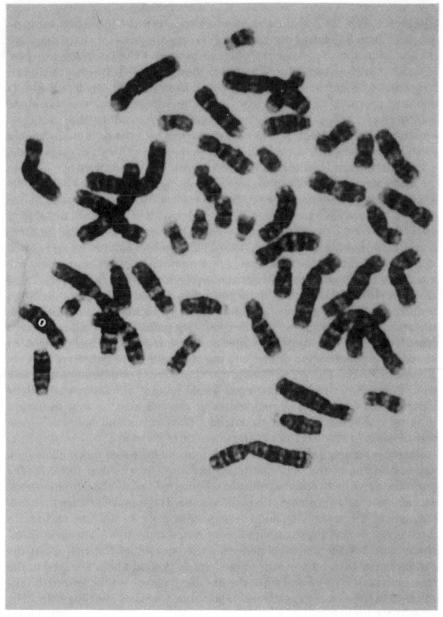

Fig. 2.1   Human karyotype at metaphase of mitosis

preparations of the cells arrested halfway through a cell division. After staining such a preparation one can count the number of chromosomes and match up the actual pairs that have come from each parent. This cytogenetic technique is further explained in Chapter 8, for it is useful in the diagnosis of certain genetical disorders in which the chromosomes are faulty. Count the chromosomes in Fig. 2.1. Is the cell haploid or diploid?

2.2 Mitosis

Mitosis is a type of cell division in which one cell produces two identical daughter cells. It involves the division of chromosomes, to give each daughter cell an identical set, followed by a physical division of the cytoplasm into two. Mitosis is the cell division of growth. It also allows each cell of the body to have a complete set of genetic information derived from the first zygote. Classically mitosis is described as being in four phases; **prophase**, **metaphase**, **anaphase** and **telophase**. In fast growing tissue these stages are accomplished in succession in less than an hour; one mitotic stage runs into the next finally ending with the division of the cytoplasm into two, a process known as **cytokinesis**.

In biology much emphasis is placed on mitosis for it is a stage when one can see with a light microscope what is actually going on in the cell. Functionally mitosis serves to separate large copied packages of DNA into two equal halves before a cell division. Look again at Fig. 2.1 which shows a cell arrested at metaphase. For a new cell to contain a complete genetic set of coded information the chromosomes must each have divided in half lengthwise and passed identical DNA molecules to each new cell. Examination of this photograph will show that the chromosomes already consist of two such identical halves. These are called **chromatids**. Each one is made up of an immensely long strand of DNA bound up with small proteins. The chromatids are identical in their DNA base sequence, in their length and in their banding pattern. They adhere to each other loosely but are physically united only at one point, the **centromere**. At anaphase the centromere divides and the chromatids are pulled apart in the division of the cell. Fig. 2.2 shows the essential steps of mitotic division for just four chromosomes. A full human cell would require 46 chromosomes to be shown. Each of the four chromosomes in the cell may be seen to have a lengthwise division into two chromatids. Each chromatid becomes a new chromosome in its own right at the moment of division.

During **prophase**, the cell's nuclear membrane is broken down allowing a free movement of the very shortened chromosomes within the cell. The centromeres, which are the waist-like constrictions to the chromosomes, become attached to protein fibrils of the **spindle**. The spindle is both a marshalling system for assembling the chromosomes prior to division and also a mechanical device for organising chromosome separation. The term **metaphase** describes the assembly period at the equator of the cell, when the chromosomes take up positions in one plane. Protein fibrils attached to the centromeres tighten up and when the spindle structure is fully formed the cell enters **anaphase** as the centromeres split under tension from the poles. The paired chromatids separate from each other and become single chromosomes at the opposite poles of the cell. The spindle fibrils not only pull these apart, but also provide a central skeletal axis pushing the poles away from each other. During **telophase** the nuclear membrane is reformed and the cytoplasm is cleft in two by a constriction of the cell. This cleavage or squeezing in half is achieved by encircling proteins, inside the membrane, that pull it in, like the draw-string on a bag.

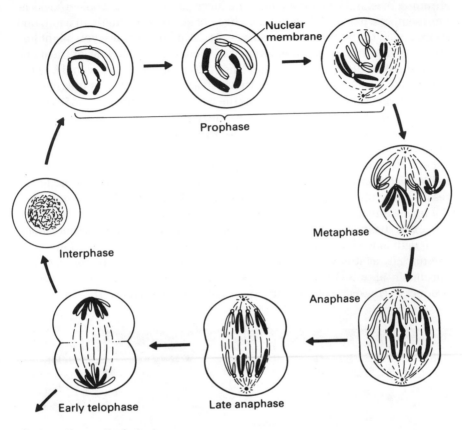

Fig. 2.2   The stages of mitosis

Mitosis is a frequent event in actively growing embryos and in replacement tissues such as epithelia. Here it takes about one hour to complete but is slower in adult tissues. The actual mitotic division is a dramatic event but one may be misled by this apparent activity into thinking that in the intervening hours and days between divisions the cell is quiet. Between divisions, at **interphase**, the chromosomes are uncoiled and less visible. During this time they are fully engaged in the transcription of the genetic code, for its translation into protein 'meaning', and in making new copies of the DNA.

2.3   The organisation of chromosomes

Chromosomes are best regarded as packaged DNA, for they only go into this tight and compact form when the cell is undergoing the trauma of a physical division into two. A typical chromosome before a cell division is best envisaged as two parallel strands of equal length with similar shape and banding patterns. Each strand is a chromatid. Where the two strands are joined to each other the junction is called a centromere. It requires some imagination to perceive the scale of the difference between minute DNA threads and the vastly larger chromosomes. Imagine for a moment a large pair of striped woollen socks as representing a pair of chromosomes. They are identical in length, in width and shape and in the position of their stripes. At 20 metres distance you might just about see that they were a matching pair. Were you then to look at them one thousand times nearer, with a hand lens at 20 millimetres you would find the fibres of the wool twisted about themselves in a thread and then knitted in loops into the larger structure of the sock: as the wool fibre is to the sock, so DNA is to a chromosome. (The very best electronmicrographs of chromosomes show them as woolly structures and not as the smooth sausages of text book diagrams.)

Each chromatid, a longitudinal chromosome half, has a diameter of about half a micrometre ($0.5 \, \mu\text{m}$ or $0.5 \times 10^{-6} \, \text{m}$) at metaphase; this is close to the resolution of the light microscope, for increased magnification will not make the image any clearer. At the greater resolving power of the electron microscope this structure can be seen to be made up of coils of **chromatin** fibre along the length of which are finer loops. Chromatin is the dark staining material of the nucleus as it is seen in fine section under the electron microscope. The chromatin fibre is 30 nanometres wide ($1 \, \text{nm} = 1 \times 10^{-9} \, \text{m}$) and is itself a product of stacked bead-like structures called **nucleosomes**. These are com-

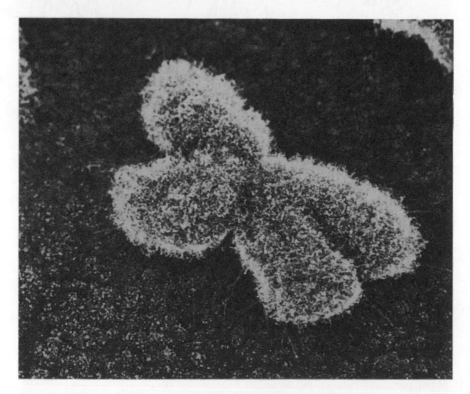

Fig. 2.3 Scanning electron micrograph of a chromosome at metaphase. The two chromatids, each made of chomatin fibres, are united at the centromere in the middle.

posed of a protein, **histone**, around which the DNA molecule itself is coiled. Each human chromatid is believed to consist of just one long DNA molecule about 5 cm in length, which at metaphase is wrapped, stacked, looped and coiled into a 5 μm length! This represents a 10 000 fold shortening. (See Fig. 2.4.) It will be clear that in this form there is no possibility of replication of the DNA or of easy transcription into RNA code. These events occur between mitotic divisions when the chromosomes are unwound at interphase of the cell cycle.

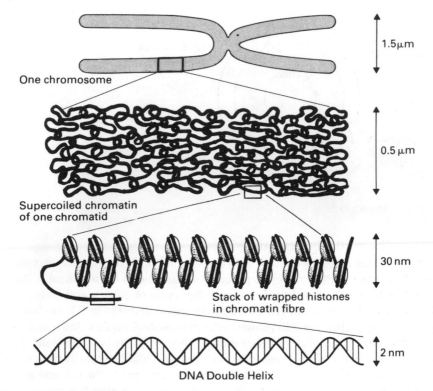

One chromosome

1.5 μm

Supercoiled chromatin of one chromatid

0.5 μm

Stack of wrapped histones in chromatin fibre

30 nm

DNA Double Helix

2 nm

Fig. 2.4   The assembly of a chromosome from DNA. DNA is wrapped around histone proteins which are stacked into a 30 nm chromatin fibre. This fibre is coiled and super-coiled to make a chromtid.

2.4   The cell cycle

Although the number of cells doubles at each division the events of mitosis occur again and again in a cycle. In human tissues actively dividing cells undergo mitosis about once a day, the actual division taking less than an hour. The cell cycle has four phases, mitosis (M), then a first gap in activity (G1), a phase of DNA synthesis (S), then a second gap (G2). (See Fig 2.5)

At the end of mitosis (M) the chromosomes lengthen and unwind. This transition involves despiralising and unhitching of loops from the chromosomes. The interphase chromosome is believed to be arranged like a necklace of nucleosome beads on a long DNA thread. During the first gap phase (G1) there is transcription of DNA into RNA at particular gene sites. During the synthesis phase (S) there is replication of the single chromosome to form a new partner chromatid parallel to it. During the second gap phase (G2) more transcription may occur briefly before mitosis commences once again. Throughout the cell cycle the cytoplasm will be growing in size, replicating its organelles and continuously synthesising proteins from the ribosomes. These events of interphase are the normal life of the cell.

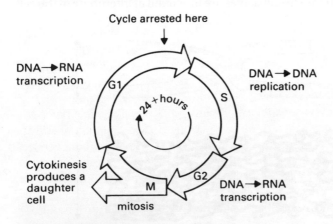

Fig. 2.5   The cell cycle

2.5   The control of cell division

Recent studies of cell cycles have shed considerable light on the genetic control of cell division itself. If the cycle proceeds to the end of G1 a critical point is passed which launches the cell into another cycle of growth and division. If however that point is not reached and the cell is arrested in G1, no further division occurs. Human cells grown in tissue culture often slow down and stop dividing after a number of generations. They seem to be genetically pro-grammed to stop. Cultured cells from older people divide less rapidly than those from children. The controls to growth are both internal to the cell and in the outer tissue environment. Skin cells for example, when grown in culture for skin grafting, inhibit each other's growth when they touch together across the culture medium surface.

If controls in the cell cycle break down, cells sometimes divide repeatedly and in an uncontrolled and invasive manner to form the tumours of a **cancer**. It is now known that in some cases the cancer may be caused by an outside viral agent. Such viruses are able to introduce a cancer gene (oncogene) into the cell. An environmentally induced change in a normal gene may also make it into a cancer gene if its code becomes so different that its protein products, which formerly exerted control on cell division, are no longer effective. An environ-mental agent that induces such a change is called a **carcinogen**. Any 'cure for cancer' will be found to operate at this level of control in the genetics of the cell. The chemotherapy drugs used in cancer treatment stop all mitotic divisions in the body. The side effects of such drugs stop of the growth of skin and hair and all other dividing tissues.

18

2.6   The replication of DNA

During the S phase DNA is replicated by complementary base pairing. This is achieved by a large number of helper proteins, principal of which is **DNA polymerase**. This enzyme builds free nucleotides onto the separated strands. But before this can happen the DNA helix must be opened into two strands, untwisted and held out straight. The sugar–phosphate chains may need to be cut and rejoined. These are enzyme controlled activities. Some of these highly specific proteins are the tools of the genetic engineer and in the future will undoubtedly be used to improve human health. (The human gene for insulin production has already been transferred by genetic engineering into bacterial cells for the production of insulin used in the treatment of diabetes.)

Figure 2.6 explains how it is thought that DNA replication occurs. It begins at several places along the length of the whole molecule. One chain of the two that are separated, known as the 'leading strand', is quickly copied as DNA polymerase passes along it adding free nucleotides. The 'lagging strand' is added to in the reverse direction, the short new lengths being linked up together by another enzyme. (See also Fig. 1.4.) One of the conservatively replicated strands stays bound to the old histones (nucleosome proteins) whilst the other is wound onto new histones. By the end of the S phase there are two parallel necklaces of nucleosomes ready to be coiled up into two parallel chromatids for the next cell division.

It is estimated that polymerisation errors (mistakes in copying) in DNA replication occur at a rate of about one in ten thousand nucleotide insertions. However, corrections are made by DNA polymerase for it appears to check the complementarity of both paired nucleotide bases as it travels along the strand.

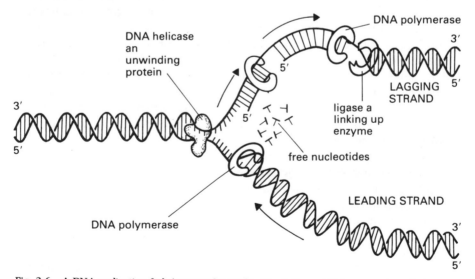

Fig. 2.6   A DNA replication fork (see text for explanation). It would take approximately two seconds for the fork to travel the length of this diagram.

19

As a result only one error in a thousand million occurs. Put in efficiency terms, DNA polymerase makes the correct complementary base pairings occur with 99.999 999 9% accuracy at an astonishing pace of 50 to 100 nucleotide additions per second. It takes about 8 hours at this rate for the *several thousand* DNA polymerases in a nucleus to copy the full human genetic code or **genome**. The odd error will almost certainly occur. Most of these will be neutral and cause no harm to their possessor but **gene mutations**, as these are called, are the source of both new variations of benefit in evolution, and sadly also the source of new genetic diseases (see Chapter 8).

2.7 Selecting the right gene: what to transcribe?

One of the mysteries of genetics is how each cell knows which proteins to make. During embryonic development different cell lines differentiate and specialise until finally at the end of the line we find a cell that is no longer **generalised** but is **specialised**, fulfilling specific roles in the tissue, within the organ and organ system of which it is a part. The nucleus may retain its full potential for developing a complete organism. If inserted into a zygote from which its own nucleus has been removed, the nucleus of a specialised cell often has the potential to initiate embryonic development all over again. Clearly such a nucleus retains a full data store, but how does it know where and when to begin decoding and in what order?

The complexity of these questions, which are still largely unanswered, is illustrated by the genes for the polypeptides that go to make up the protein **haemoglobin**. This red blood pigment is only synthesised in the final stages of the development of a red blood cell. In an adult the appropriate genes are activated in the tissue of the bone-marrow alone. Here two genes are turned on, one for the alpha (α) polypeptide chain and one for the beta (β) chain of the haemoglobin protein. Many adults have other variant polypeptides, such as delta polypeptide (δ), that are synthesised as well. In the human embryo a different, epsilon (ϵ), polypeptide is produced whilst another gene for gamma (γ) polypeptide allows the synthesis of foetal haemoglobin. Thus each of the polypeptides, α, β, γ, δ and ϵ are called into play in a person's life to match the different blood protein requirements at different ages and conditions. It is not known how this is done.

When it comes to switching genes on, two sorts of gene regulation seem to occur: there are **negative** and **positive controls**. In negative control a regulatory device stops the DNA from being copied. This **repressor** is likely to be a protein which prevents RNA polymerase from transcribing a gene. A famous example of negative control is shown in the human colon bacillus *Escherichia coli*. The bacterium can bring a milk digesting enzyme into play as and when it is needed. Supposing that lactose has been taken up from milk in the bacterial environment, it then binds to the enzyme gene repressor and unlocks it from one specific gene site. This action switches on the gene to make the mRNA that makes the appropriate enzyme, called galactosidase, for lactose digestion, inside the bacterium. This is an important adaptation for the bacterial economy, for had the lactose not been available as a food to the bacterium then making this particular digestive enzyme would have been wasteful.

Positive controls are those in which a regulatory device turns on the gene transcription directly by acting as an **activator** of RNA polymerase. In insects, where such controls have been studied, the moulting hormone causes the cells and nuclei to swell and this stimulates the transcription of the genes concerned in producing the new cuticle. When this happens large 'puffs' of chromosome unravel and RNA is made by the hormone's direct action on an **activator protein** that promotes transcription. Very little is known of gene controls in humans but they are believed to be largely of this positive kind.

2.8   Transcription and translation of the code

Genes are located singly or in small groups together on loops of chromatin on the chromosome. In a condensed and stained chromosome the bands along its length are the sites of these genes. It is here also that the chromatin unwinds in 'puffs' and the process of **transcription** occurs. After a length of DNA has been 'switched on' by some gene control, the chromatin unravels, the DNA chains part and RNA polymerase transcribes the DNA code from one of the two chains. This is the 'sense strand' for the synthesis of messenger RNA (mRNA). Transcription follows the coding rules set out in section 1.9. mRNA will be released from the chromatin loops as it is formed and then diffuses out of the nucleus to the cytoplasm. In human cells a greater length of DNA may need to be transcribed to make a full length of mRNA for translation into a single protein.

Translation from mRNA into protein requires the mediation of transfer RNA molecules (tRNA) and the ribosomal RNA molecules (rRNA) in the ribosomes. tRNA molecules are relatively small possessing a folded clover-leaf shaped structure. At one end of this 'adaptor' molecule they carry an amino acid and at the other they have a protruding triplet code, often drawn as a three pronged fork. This triplet will pair with the **codon**, or code triplet, on the messenger RNA. Faithful protein synthesis is dependent on this complementary pairing of triplets and specific matching of tRNAs to specific amino acids.

Ribosomes consist of proteins and **rRNA** locked into an assembly line device that mechanically draws the mRNA through a groove in its surface. The mRNA may be regarded as a tape of coded information. It passes through this groove with what is perhaps a rapid jerky motion, passing in the same direction in which it was transcribed when initially built on the DNA molecule. Like the electromagnetic tape in a tape-recorder its information is translated as it passes through the ribosomal 'machine' (but instead of music it produces a polypeptide chain!). The movement of the mRNA through the ribosome is limited by the rate at which tRNA molecules can align themselves on the ribosome groove. In the living cell these different tRNA molecules, with their attached amino acids, are in continual kinetic motion around the ribosome. When a tRNA molecule, with an **anticodon** complementary to the codon of the mRNA, fits into place the mRNA shifts along a triplet length and makes room for another tRNA to come in. This sideways shift must displace the previous tRNA but this will not be released from the ribosome until it has given up its amino acid. This amino acid will by then have condensed onto the amino acid just previously delivered, so forming the growing polypeptide chain, the sequence of which has been governed exactly by the mRNA passing through the ribosomal groove. When finally a **UAA**, **UAG** or **UGA** signal is reached no more amino acids are added and the synthesised protein is released.

It is important not to underestimate the translational power of a gene once it is switched on. In the glands of the silk worm caterpillar, which spins silk for its cocoon, each cell has only one silk gene, yet it can make 10 000 mRNA transcripts, *each* of which directs the synthesis of about 100 000 silk protein molecules in just four days; this represents a total of 10^9 molecules of silk from one strip of code. Similar rates of synthesis pack each of our red blood cells with 250 million molecules of haemoglobin. As red blood cells are produced at a rate of two and a half million per second your body's rate of synthesis for just this one protein is 6×10^{14} molecules per second. Protein synthesis may be on a small scale but it is big business.

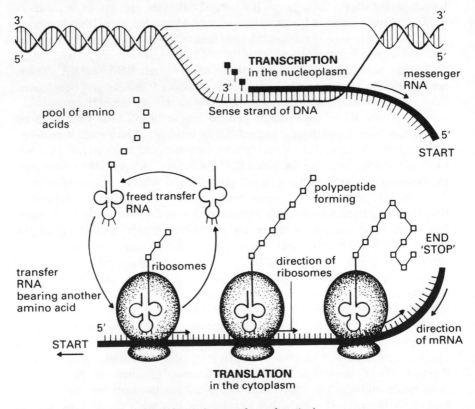

Fig. 2.7   Transcription and translation (see text for explanation)

2.9   The differentiation of cells

In the human embryo all the cells have the potential to form a complete individual up to about the 8-cell stage. Before this point separated cells can become identical siblings (twins, quadruplets, etc.). From this early stage the cells then begin to develop differences of character according to their physical position in the embryo and the mutual environment that they create for each other. Organiser substances produced in one region may establish gradients of influence which govern the way that other cells subsequently behave. Cells become different by dividing unequally and seemingly making choices, or having their choices directed, as to which path they should take.

Figure 2.8 shows how a cell whose function in life is to produce the hormone insulin, has specialised. It will have arrived by a pathway in which genes have undoubtedly made the rules, and events will have taken an interactive yet largely predetermined course. At the end of the mystery of human development we find millions and millions of cells, each with an allotted function in a highly integrated whole.

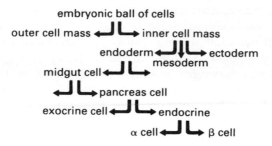

Fig. 2.8.

3 Genes in the organism: meiosis, Mendelism and alleles

3.1 Introduction

So far we have examined the nature of genes and seen how their coded instructions find expression in the proteins of cells. At first sight it is hard to link this idea with the much wider one of genetic inheritance. That is the aim of this chapter.

The coding of faithfully copied DNA may explain why children resemble their parents but it does not explain why brothers and sisters are so different from each other. It is important to understand the way in which sexual reproduction generates new combinations of genetic material. This is achieved through **meiosis**, the cell division that forms **gametes**, and by the new combination of characteristics that follows fertilisation. The mathematical elegance of inheritance patterns was discovered by **Gregor Mendel** who, on the basis of his results, perceived that hereditary characters occurred in pairs. **Alleles** are forms of a gene found in an individual organism in pairs.

3.2 The evolution of sex

The existence of male ($\male$) and female ($\female$) individuals is widespread in both the animal and the plant kingdoms. Even where there are advantages in one individual being both male and female at the same time, that is hermaphrodite ($\hermaphrodite$), the evidence all points to benefits in sexuality itself that are not to be found in less complicated asexual ways of reproducing. Reproduction without sexuality may be very simple, but there are few higher plants and animals with *only* **asexual** processes of reproduction. Asexuality by definition involves reproduction without sex cells and must be achieved by growth and mitotic divisions alone. Such a population of genetically identical individuals is called a **clone**. Clonal offspring have short term advantages to their parents, allowing rapid reproduction of a genetically uniform type, for example in the summer populations of aphids. Clones lack genetic variability because it is only faithful mitotic growth divisions that produce the new individuals; they are all genetically the same. There is much science fiction talk of human 'clones'. It may one day be technically possible, but is certainly undesirable, to produce large clones of people! Because of the way that they arise identical twins are in fact a clone of two. One of the distinctive features of people that we often forget is that they are all so uniquely different. This is a direct result of the biological way that we reproduce.

What is the origin of sex? It seems that the very earliest cells evolved a primitive form of sexuality in which one cell donated short lengths of DNA to another. Amongst bacteria this is sufficient to generate more variation than mutation alone allows. The earliest single-celled animals and plants with true chromosomes may have gained some benefit from obtaining another nucleus from an individual of the same species. Any cell with one set of chromosomes is termed **haploid**. What therefore began as one haploid cell engulfing another to obtain its nucleus may be paralleled today in the events at fertilisation. There might have been immediate advantage for these early cells in doubling up their chromosomes to become **diploid**, with two gene sets. Such an arrangement might have guarded against mutation having damaged the single copy of a gene or might have contributed to valuable variation in the new possessor. But managing two sets of chromosomes is complicated. Very early in evolution such forms of life seem to have developed a system of aligning two similar chromosomes, from the contributing cells, side by side, perhaps to make possible the removal of mistakes in the coding of one by a process of copying or exchanging code lengths from the other. Also, by *exactly* separating such pairs of chromosomes in a cell division, which halved the total number again, a cell might reassemble a complete haploid set. Side by side exchange and exact halving of diploid chromosome sets is what we see today in meiosis. This is the cell division of sex cell formation that produces haploid eggs and sperms. As we shall see in this chapter, such divisions produce new haploid cells that are not identical. Before this point in evolution all variation was due to mutation alone. The new ability to generate more variation may have had crucial evolutionary advantages, for meiosis and fertilisation are found today throughout the plant and animal kingdoms.

To summarise:

- two sets of chromosomes have survival advantages over one set;
- the re-formation of sex cells with single sets produces variation;
- fertilisation produces new combinations of the variety;
- such variability within a species contributes to species survival, for sexually reproduced individuals are all different from each other.

We believe that this is the basis for the evolution of sexual reproduction as we find it in ourselves; variety is the spice of life.

3.3   The logic of meiosis

Meiosis or **reduction division** involves an exact halving of chromosome numbers. Fertilisation restores that number by doubling the halved number, as the **gametes** or **sex cells** fuse. The diploid state $(2n)$ is one in which the chromosomes are present in pairs. These pair members are often described as **paternal** and **maternal chromosomes** for they originate from the male and female parental gametes respectively. Such chromosomes are said to be **homologous** because they are identical in shape, size, banding-patterns and position of their centromere. Each gene is therefore represented twice in each cell, once on one position on one chromosome and again at the same position on the other chromosome of the pair. Each representative of the gene is termed an **allele**. (These two versions of the same gene give the individual an insurance policy, for it is quite possible for one to be damaged by a mutational change.) After fertilisation the new combination of chromosomes will give the zygote and new organism a unique **genotype**. In order to reproduce another generation by sexual reproduction the genetic material must be halved again to form new gametes. It is not sufficient to do this randomly, for each of the chromosomes of a pair carries the same sequence of genes. Just to halve the total number of chromosomes in a disorganised manner would leave the daughter cells with incomplete sets of genes. However, if each maternal and paternal pair are precisely separated then the gametes will each have a full haploid set of genetic information to pass on. When these gametes come together at fertilisation in a new combination the chromosomes will pair up and each gene will have two representatives. The full diploid ($2n$) state will be restored.

3.4   The stages of meiosis

Fig. 3.1 outlines the main stages of meiosis for a diploid cell with four chromosomes. In these diagrams $2n = 4$, so the number of homologous pairs is 2, i.e. $n = 2$. For a human cell undergoing meiosis $2n = 46$, so the pattern shown here is very greatly simplified when compared to the complex events in a human reproductive cell.

Meiosis consists of one cell undergoing two divisions to make four cells. The stages are, as in mitosis, **prophase**, **metaphase**, **anaphase** and **telophase**. The two divisions are numbered I and II respectively. It will be seen that there is no interphase between telophase I and prophase II. It is important to recognise the significance of this fact. As there is no DNA replication and no new chromatid formation between the two divisions, the result must be a halving of the genetic material. The change from being **diploid** to **haploid** occurs during the first division; at anaphase I the homologous chromosomes are separated and the 'reduction division' takes place. The second division involves the separation of the two chromatids in the chromosomes and is no different from an ordinary cell division.

The key events of meiosis occur in **prophase I** and it is this stage which differs so markedly from mitosis, with which it should be contrasted. Prophase I begins in a similar way to mitosis but the paternal and maternal chromosomes, instead of behaving independently, seek each other out and come to align themselves, side by side, in their homologous pairs. (In Figs. 3.1 and 3.2 they are given different shades in the diagram to show their paternal and maternal origins.) In these positions the individual chromosomes, each with their two chromatids, soon form four aligned strands. By the end of prophase I these four chromatids are so closely bound and inter-twined that the united pair of chromosomes is often called a **bivalent** (worth two) to distinguish it from a single chromosome. There then follow two key events: **crossing over**, in which chromatids exchange lengths, and **independent assortment**, in which either of one pair of chromosomes may go to the poles of the cell with either of any other pair in a seemingly random manner. Crossing over and independent assortment are very important events in generating **variation**. These two processes within meiosis I are described in the next section.

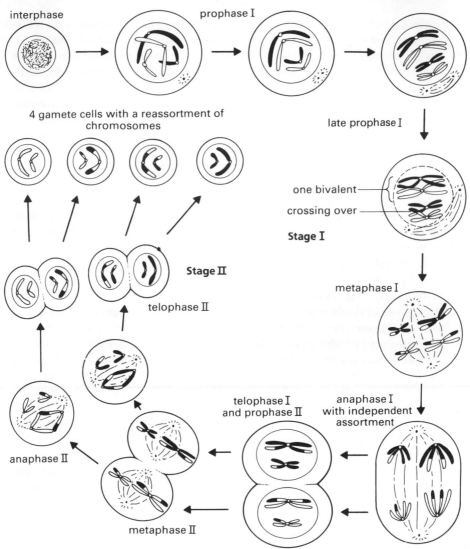

Fig. 3.1   Meiosis (for explanation see text)

3.5   Independent assortment

The shading of the chromosomes in Fig. 3.1 serves to indicate the paternal and maternal origin of these chromosomes. In the figure each bivalent is seen to be formed from two pairs of chromatids each of different shade but there is no evidence that the cell knows which chromatids are maternal in origin and which paternal; in other words, the orientation of the maternal and paternal chromosomes at metaphase I is entirely fortuitous. Each gamete will therefore receive either a maternal or a paternal chromosome from each bivalent that divides; because the orientation of the pairs are at random, each gamete will be given some of each parental type. This independent assortment of homologous chromosomes leads to a reassortment of homologous chromosomes between cells.

27

It is helpful to put this in personal terms. The egg which originally gave rise to you itself contained an assortment of chromosomes from both of your grandparents on your mother's side of the family. The sperm that fertilised the egg that gave rise to you contained an assortment of chromosomes from the grandparents on your father's side of the family. You have therefore, exactly half of your parents' chromosomes and approximately a quarter of each of your grandparents' chromosomes. Your brothers and sisters will also have exactly half of their chromosomes from each of your parents, but because chromosomes segregate randomly at each meiosis they are not likely to be the same half portion as your own received half, nor yet entirely different. On average you will share half of your genetic material with your brothers and sisters and one quarter with your grandparents. This is only an average and you may well share a little more or less. The cellular events of meiosis and fertilisation go some way to explaining family resemblances and also the possibility of a less than equal inheritance of **traits**, or hereditary characters, from our grandparents and great-grandparents.

3.6   Crossing over

The degree of shuffling of hereditary characters is greater than random segregation of chromosomes alone would allow. The reason for this is the formation of cross-overs or **chiasmata** between chromatids in a bivalent. Chromatid exchange, illustrated in Fig. 3.2, means that each chromosome is in fact only partly maternal and partly paternal after crossing over has taken place. A chromosome with a paternal centromere may in fact, after crossing over, carry more maternal than paternal DNA, but generally this is not the case. Crossing over increases the variability of gametes and increases the equality of representation in an individual of the genes from each of their four grandparents. The actual events of crossing over are not well understood, but it is clear that chromatids may break and cross-join relatively easily at the most condensed stage in the middle of prophase. The term 'crossing over' is slightly confusing, for no twisting over of chromatids occurs, merely a disconnection of two adjacent chromatid strands followed by a cross connection. A cross-over forms a shape representing the Greek letter χ (chi, pronounced *kie*) and hence is also called a **chiasma** (plural: chiasmata). Chiasmata may take place between any of the four chromatids of a bivalent. In humans it is common for from one to three

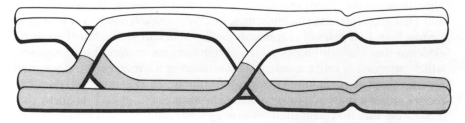

Fig. 3.2   Chiasma formation. Paternal and maternal chromosomes are shaded differently. Two crossovers are shown involving exchanges in three out of four chromatids in one (bivalent) chromosome pair.

cross-overs to occur between homologous pairs at prophase I. In summary, crossing over increases the variability of gametes. This exchange is called **recombination**.

3.7 Mendelism

A contemporary understanding of human genetics has many strands leading to it. The discovery of nucleic acid coding and the behaviour of chromosomes at cell division was very important, but long before this knowledge was available key ideas about the pattern of inheritance in plants and animals had developed from quite different types of study. In this Gregor Mendel (1822–1884) is rightly regarded as the founding father, for his elegant experiments on inheritance in peas (*Pisum sativum*) made clear the particulate nature of hereditary characteristics. Mendel was a well-educated monk with responsibility for the vegetable garden of his monastery (in a part of Europe at that time in the Austrian Empire). It was because of Mendel's work that the science of genetics got off to such a good start in 1900. Only then did the events of meiosis become related to his discoveries. Mendel knew little of cellular processes and nothing of meiosis, yet he was able to work out the essentially paired nature of hereditary characteristics and the phenomena of genetic dominance and recessiveness. The fine details of his experiments are commonly set out in advanced biology text books. It will suffice here to set out the logic of his conclusions from the briefest glimpse of his work in the monastery garden at Brno.

Mendel chose a plant species with established variety and well-defined features. He confined his attention to the inheritance of seven pairs of apparently contrasting character, such as tall and short stem, red and white flowers, etc. As peas are normally self-pollinating they are typically uniform in genetic nature and 'breed true'. Thus when Mendel, crossed together different lines of pea plant and followed the inheritance pattern over the next four generations, he was able to trace the reappearance of the characters he had mixed together. His university training in mathematics perhaps gave him an awareness of the ratios that he obtained. From their significance he tumbled to the important idea of **hereditary particles** being in pairs.

P	Parental generation	round	crossed with	wrinkled
F1	First filial generation		all round (self fertilised)	
F2	Second filial generation	round	and	wrinkled
	Numbers obtained	336		101
	Approximate ratio	3		1

Taking just the example of the cross between peas with round and wrinkled seeds, we can follow his thinking. The appearance of the seeds in each generation and their numerical proportions in the offspring are shown.

In the **first filial generation** (of brothers and sisters) the round character prevailed, but in the second, although round was the most common, about one quarter were wrinkled, like one of the original parents. On this basis Mendel argued that the round pea character was **dominant** to the wrinkled pea, which in the first filial generation (F.1) was clearly the **recessive** character. The suppression was not permanent, for in the **second filial generation** (F.2) the 'wrinkled' character reappeared once again.

Mendel argued that the 'wrinkled' character had not disappeared in the F.1 only to reappear spontaneously again, but that it had been hidden; it could be regarded as a discrete particle that had neither been diluted nor lost but was still present in an unexpressed form. Mendel argued that this 'wrinkled' character must be present alongside the 'round' one in the F.1 if it was to reappear again in the F.2. If this was the case then logically each plant had to contain at least one contributory factor from each of its parents. He reasoned that if the F.1 individuals had one such character from each parent, then those characters should be in pairs. If this was true for this generation could not all generations have their characteristics in pairs? Mendel then tested this idea. He represented the characters as letters, with *AA* for the round-seeded pure breeding parent plant (with capital letters for the dominant type) and *aa* for the wrinkled-seeded pure breeding parent plant. If each of these two parents donated one of their particles to the F.1 it would then possess *Aa*, and be round seeded because of the dominance of this particle. The elegance of this was that the F.1 would then pass on either one or the other of these two in a single gamete to form the next generation. This would then result in the three to one ratio he had observed, one quarter of the F.2 being of the recessive wrinkled-seeded type.

Parents	round	X	wrinkled
	AA		*aa*

Gametes	*A*		*a*

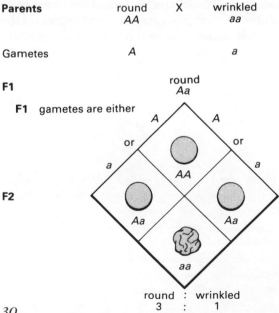

F1 — round *Aa*

F1 gametes are either

A or *A*

a or *a*

F2

AA

Aa — *Aa*

aa

round : wrinkled

3 : 1

Before he arrived at this conclusion Mendel had carried out other crosses between different pure-breeding strains and obtained similar ratios each time. This established what has come to be known at **Mendel's First Law**:

'of a pair of contrasting characters only one may be passed on in a single gamete'.

The pertinence of this to meiosis should now be clear. Mendel was expressing by this discovery what we now see to be the reality of homologous chromosomes separating from each other in the formation of gametes at meiosis.

Mendel followed these **monohybrid** crosses with **dihybrid** crosses where the expression of two pairs of characters was studied through three generations. The example of yellow- and green-seeded plants together with round- and wrinkled-seeded plants is well known. In this case the yellow character was dominant for it was the only seed colour in the first generation (F.1). In the second generation (F.2) it was present in three-quarters of the offspring, approximately one quarter being green. Thus when pure-breeding round- and yellow-seeded plants were crossed with green- and wrinkled-seeded plants (which were doubly recessive) each pair of characters (e.g. round and wrinkled) behaved quite independently of the other pair (e.g. yellow and green). In the F.2 therefore, a second 3:1 ratio seemed to overlie the first 3:1 ratio and so produce a more complicated looking ratio, between the four F.2 offspring types, of approximately 9:3:3:1.

P	(Parental generation)	round yellow	X		green wrinkled
F1	(First filial generation)		all round yellow		

F2	(Second filial generation)	round yellow	round green	wrinkled yellow	wrinkled green
	Numbers obtained	315	108	101	32
	Approximate ratio	9 :	3 :	3 :	1

Mendel sought to explain this ratio by postulating once again not only a clear dominance (round and yellow) and recessiveness (wrinkled and green) but also an independence of the single characters, as represented by letters, from one another. In the formation of the gametes either A or a could go together with either B or b. This meant that the F.2 were producing four types of possible gametes, AB, Ab, aB and ab. As these plants were being self-pollinated, gametes would come together as a square of the four gamete combinations, making sixteen offspring combinations in all. $(4 \times 4 = 16)$; $(9 + 3 + 3 + 1 = 16)$.

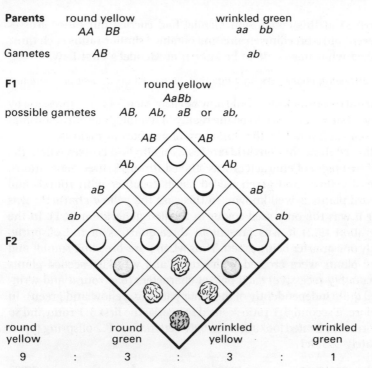

Parents   round yellow   wrinkled green
             AA BB   *aa bb*

Gametes   *AB*   *ab*

F1   round yellow
                 AaBb

possible gametes   *AB, Ab, aB* or *ab,*

F2

round yellow   :   round green   :   wrinkled yellow   :   wrinkled green

    9   :   3   :   3   :   1

This led Mendel to formulate a second idea:

'of two pairs of contrasting characters either of one pair may go together with either of another pair in the formation of a single gamete'.

This is often referred to as **Mendel's Second Law**. Perhaps Mendel was lucky to find this out for we can now see, with the benefit of hindsight, that he had selected a second pair of characters on a different pair of homologous chromosomes from the first pair (see Fig. 3.7). Today we would express his discovery by saying that either one of a pair of homologous chromosomes may go to the same pole of the cell as either of another pair of homologous chromosomes at the first division of meiosis; homologous chromosomes assort independently of each other.

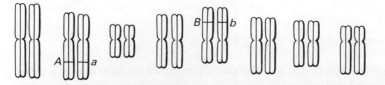

Fig. 3.7   The seven pairs of chromosomes in the pea showing the position of two pairs of independently assorting alleles on separate chromosome pairs.

We are indebted to Mendel for *four things*: a clear concept of genes as particles, the idea of dominance and recessiveness, the perception of the occurrence of characters in pairs in the adult and as single copies in the gamete, and the notion that such characters may segregate freely from each other in forming the offspring of the next generation. It is a shame that this pioneering work went unacknowledged in his lifetime. But it is to his credit that he wrote up his researches in detail. Mendel was promoted to responsibilities away from his beloved peas; after concluding his researches he became the abbot of the monastery at Brno, today a town in Czechoslovakia, where the walled monastery garden may still be visited.

3.8 The concept of the gene

When Mendel did his work chromosomes and genes were unheard of. It was not until 1877 that Walther Flemming first decribed the chromosomes within the nucleus; ten years later Weissmann described meiosis as being different from mitosis. Sixteen years after Mendel's death, in 1900, his work was rediscovered by researchers looking for other published evidence of patterns in the hereditary process. In 1902, quite independently, Sutton and Boveri saw the events of meiosis as acting out, at the cellular level, what Mendel had described as the independent assortment of hereditary characteristics. The term 'gene' was coined to describe these paired entities in the diploid cell, one member of the pair being passed on in a single gamete.

In the early part of this book the term 'gene' has been used loosely for that length of DNA corresponding to a strip of mRNA that codes for the synthesis of a single polypeptide or protein. This is the **functional gene**. By contrast, Mendelian genetics has grown up to see the gene as being a component of a chromosome with a particular **locus**, or position, on a pair of homologous chromosomes. Chiasmata (cross-overs) take place between and not within such a gene locus. This 'bead on a string' concept is the **structural gene**. To a great extent the two ideas are compatible, although many human genes are not continuous DNA lengths, but in fact exist in a sequence of strips often close by each other. These may be united in one messenger RNA, or combine in their effects, when separate polypeptides unite to make one functional protein. Mendelism has also shown us that the gene is a paired structure, for it exists on two homologous chromosomes, and may well do so in more than one form.

Each chromosome's gene form is called an **allele**. One individual person has two alleles at each locus, one on each of a pair of homologous chromosomes. These two alleles may have exactly the same code or they may be different. Because the coding of alleles differs, they may have different effectiveness in the life of the cell. This is what is meant by the relative 'dominance' and 'recessiveness' of gene expression. It was Mendel's perception of units with differing status that led him to use capital letters of the alphabet for dominant characters and small letters for recessive ones. (In this chapter Mendel's original lettering for the round and wrinkled, and yellow and green characters has been used.) Today we still use letters to describe the genetic constitution or **genotype**. When this is done the locus 'A', which is the gene, is represented by allelic italic letters A or a, for the dominant and recessive alleles respectively. The next chapter sets out to unite the classical idea of the Mendelian character with the modern biochemical idea of the functional gene. In doing so we shall take examples principally from human heredity and introduce the full vocabulary of terms that students of genetics should have.

4 The patterns of Mendelian inheritance in humans

4.1 Genetical terms

Nothing is more defeating in genetics than the language used. The most important and commonly used terms should be fully understood before discussing the simplest patterns of heredity displayed by people.

Dominance and recessiveness

Where alleles are of two types in an individual the one which expresses itself the more is said to be dominant, whilst that which is expressed the less is said to be recessive. **Complete dominance** leaves no expression of the recessive allele at all. The differing ways in which the alleles express themselves are described later in this chapter, where careful consideration is given to defining **incomplete dominance** and **codominance**. (See 4.5.)

Genotype and phenotype

The **genotype** of an individual is its genetic constitution with respect to the allelic partners in a particular gene. As we have seen in the last chapter, a pure-breeding yellow pea plant may have the assigned genotype *BB* whilst the pure-breeding green pea plant has the genotype *bb*. Mendel used alphabetical letters in his written papers and we have in fact followed his original lettering in the last chapter. It is common to use genotype symbols that are shorthand for the character in question. Thus in the example above, if yellow is dominant, we could have used *YY* for yellow and *yy* for green, describing that locus as 'Y'. Quite often the notation $+$ or $-$ is used. Here a minus sign stands for a defective allele or one lacking expression of the gene product; for example, in the Rhesus blood group a Rhesus positive person has at least one Rh^+ allele at the 'Rh' locus and may have the genotype Rh^+Rh^+ or Rh^+Rh^-. It is also a common practice to use a superscript to distinguish variant alleles from each other, for example individuals of blood group AB have the genotype I^AI^B.

Wherever a pair of genotype symbols appear for an individual it can be assumed that they will separate at meiosis in gamete formation. Thus diploid individuals will have two-letter genotypes, for one particular characteristic, whilst the gametes will have only one. The **phenotype** of an individual is their observable appearance with respect to the hereditary character under consideration. Although the phenotype expresses the concealed genotype it is also influenced by the surrounding environment; one should not assume therefore that the phenotype is due to the genotype alone. (See Chapter 9.)

Homozygotes and heterozygotes

For any genetic constitution the alleles received from the two parents may be the same (homo-) or different (hetero-) when they first come together in the zygote. **Homozygous** individuals may be of either the dominant or recessive genotype e.g. *AA* or *aa*. **Heterozygous** individuals have a mixed genotype, e.g. *Aa*, and the phenotype will be due to the expression of the dominant allele. Individuals of the two types are called **homozygotes** or **heterozygotes** accordingly. Individuals homozygous for one locus may of course be heterozygous at another. Humans are perhaps homozygous at 90% of their gene loci. More homozygosity is likely to be the result of in-breeding. A high degree of heterozygosity is a product of out-breeding. The cross-breeding of distantly related individuals within a species may lead to an advantageous genetic position resulting in greater fertility, greater size and increased disease resistance. This is known as **heterosis** or 'hybrid vigour' (see 10.2).

4.2   Chance: probability and ratios

One of the striking features of Mendel's work was the whole-number ratios that he obtained. Such mathematical ratios can only be obtained with any accuracy from a large sample. Their pattern taught Mendel the rules that govern inheritance; in this sense ratios reflect the facts of inheritance and do not govern the events. This can be illustrated by a simple example. If you toss a coin it will come down as heads or as tails. Whether it falls as heads or as tails on any occasion is not governed by which way up it landed on the last or any previous occasion. To assert with confidence that the coin has an equal chance of coming down as heads or as tails one would need to toss it more than a hundred times and then apply a suitable statistical test to the result.

Chromosomes segregate randomly to either end of the cell in the first division of meiosis. Like the toss of a coin it is a 50:50 chance which way a maternal or paternal partner chromosome goes. If one such chromosome pair carries a gene represented heterozygously by alleles *A* and *a*, there will be a half chance (a probability of 0.5) that a single gamete will contain the *A* allele and a half chance that it will contain *a*. For another pair of alleles on another pair of chromosomes, *B* and *b*, there is again a half chance of *B* or *b* being in each gamete. As each pair of chromosomes separates to either pole of the cell at meiosis, at the end of the first division, either of the first pair may go with either of the second, in making the gametic genotype. Hence there is a $\frac{1}{2} \times \frac{1}{2} = \frac{1}{4}$ chance of each of the four combinations *AB*, *Ab*, *aB*, and *ab*. (Refer back to Fig. 3.1 and Fig. 3.6 in Chapter 3.)

4.3   Pedigree charts

It is common to express inheritance patterns in human families by means of symbols in pedigree charts. Males are shown by square symbols and females by round. The shading-in of the symbols indicates phenotypic expression in each person. Pedigree charts show each generation at a single level; one reads generations in time down the page and in birth sequence from left to right across it. Pedigree charts are useful for determining the form of inheritance involved by interpretation of the pattern displayed. Once a dominant or recessive pattern is inferred some of the genotypes may be deduced and predictions as to probabilities in further generations may be made. Where the pattern of inheritance correlates reliably with the sex of the individuals, sex-linked inheritance patterns may be suggested (see Chapter 5). Fig. 4.1 shows a pattern for the inheritance of a dominant allele. It never skips a generation and possessors always have one parent with the characteristic. Fig. 4.2 shows an inheritance pattern for a recessive allele. Here expression only occurs in the homozygous recessive condition. First cousin marriages often reveal the presence of a recessive trait that last appeared in a common grandparent.

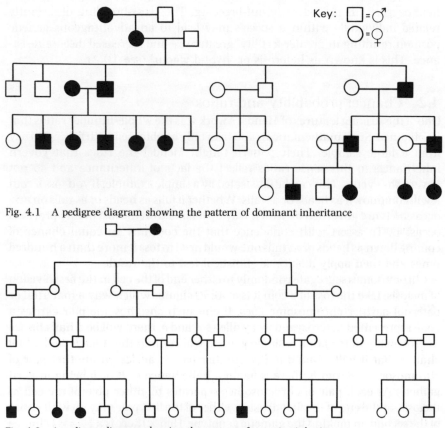

Fig. 4.1 A pedigree diagram showing the pattern of dominant inheritance

Fig. 4.2 A pedigree diagram showing the pattern of recessive inheritance

4.4 Monohybrid inheritance in humans: PTC tasting

Phenylthiocarbamide (PTC) is an organic chemical that for most people has a bitter taste even when very dilute. The threshold at which different people are able to taste it in solution falls either side of an intermediate concentration. It shows in this distribution (Fig. 4.3) what is classically called a **discontinuous variation**.

If a 0.01% solution of PTC is prepared, those tested for their ability to taste it fall into two discrete groups with very few on the border line. About 30% of people in the U.K. at this threshold level cannot taste anything unpleasant, whilst the remaining 70% find it bitter or extremely so. (This biochemical is known to be

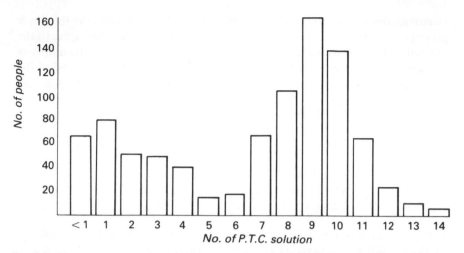

Fig. 4.3 Discontinuous variation in PTC tasting (see text). Solution 1 is 1.3 g dm⁻³ strength (1300 parts per million). Solution 2 is half this strength. Solution 3 is half again, and so on down the series to 14. The 'non-tasters' (left) are only able to taste the chemical at strength. Some 'tasters' (right) can detect as little as one part per million. Different populations show slightly differing thresholds indicating the probable influence of modifying alleles. (After Kalmus.)

carcinogenic in large doses and it is related to bitter compounds naturally found in plants. There may therefore be adaptive advantages in being able to taste it, though on the other hand being repelled by it may limit one's food supply. Interestingly this same genetic characteristic is found in other higher primates besides ourselves.)

The two groups of people are called 'tasters' and 'non-tasters' respectively and the ability to taste the compound is very largely governed by a single gene. Evidence for this comes from the following. Marriages between tasters more often than not produce a large majority of children that are tasters but a few that are not, whilst marriages between non-tasters always produce non-taster children. All tasters have at least one taster parent. This leads us to see that the single gene locus should be called 'T', after this tasting ability, and that the phenotype 'taster' is due to the possession of a dominant allele, which we shall call T. The genotype of tasters is thus TT, if homozygous, and Tt, if heterozygous, whilst the phenotype of 'non-taster' has the genotype tt. Following through the inheritance patterns for this single genetic character will help to reinforce an understanding of the simplest patterns of heredity and introduce, to those unfamiliar with them, the correct use of terms and best layout of the page in tackling genetic problems.

As only one of a pair of homologous chromosomes may be present in a single gamete, each egg or sperm only carries a single allele of the parental pair.

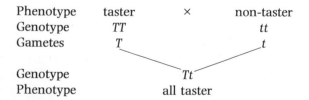

Phenotype	taster	×	non-taster
Genotype	TT		tt
Gametes	T		t

Genotype   Tt

Phenotype   all taster

Here only one gamete type is possible from each parent. If however the taster parent is heterozygous that parent will produce two gamete genotypes. In the example below, because the non-taster is homozygous recessive, the nature of the gamete from the heterozygous parent will govern the phenotype of the offspring. This 'either/or' situation is neatly illustrated in the 1:1 ratio obtained.

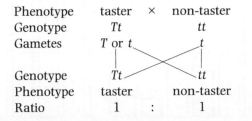

Phenotype	taster	×	non-taster
Genotype	*Tt*		*tt*
Gametes	*T* or *t*		*t*
Genotype	*Tt*		*tt*
Phenotype	taster		non-taster
Ratio	1	:	1

In the last example it is important to remember that two such parents will not produce children in an exactly 1:1 ratio any more than parents will have boys and girls in equal numbers.

If both parents are tasters a number of different situations may arise. If both are homozygous *TT*, then clearly each will have one gamete genotype and all the offspring will be *TT* like their parents. If however one of the parents is heterozygous, that parent will produce two gametic genotypes.

Phenotype	taster	×	taster
Genotype	*TT*		*Tt*
Gametes	*T*		*T* or *t*
Genotype	*TT*	*Tt*	
Phenotype	all taster		

Here two genotypes occur in the offspring but only one phenotype. If both of the parents are heterozygous each will produce two types of gamete resulting in approximately one quarter of the offspring being of the recessive phenotype.

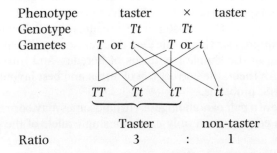

Phenotype	taster	×	taster
Genotype	*Tt*		*Tt*
Gametes	*T* or *t*	*T* or *t*	
	TT *Tt* *tT*	*tt*	
	Taster	non-taster	
Ratio	3	:	1

This is the classic Mendelian F.2 ratio, of 3:1, that arises from an F.1 cross of two heterozygotes. For each child born to these heterozygous parents there is a chance of 1 in 4 that a non-taster will occur. This 3:1 ratio does not mean that of four children born one *must* be a non-taster!

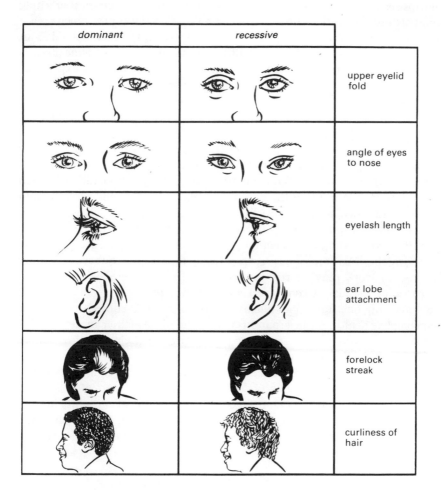

Fig. 4.4   Some phenotypic differences in humans attributed to single gene effects. (After Rostand and Tetry.)

4.5   The expression of alleles

Human genetics is peculiarly difficult because there are so few external features of people that can be relied upon to be attributable to a **single gene effect**. Fig. 4.4 gives some possible examples, but all are rather suspect if examined critically! Even the classic 'blue eyes' being recessive to 'brown eyes', that is often quoted in elementary genetic texts, is an oversimplification. Too few of these features are clearly 'discrete variables', that is are either one thing or the other, reflecting simple dominance and recessiveness. If we return to the definition of a gene, as that length of DNA responsible for encoding a single functional protein, we may look for many neater examples than these of single gene effects in human metabolism. Some of these can help us to understand the nature of dominance much better.

The disease galactosaemia: complete or incomplete dominance?

Galactose is a monosaccharide sugar received by babies in their mother's milk as part of the disaccharide milk-sugar lactose. Galactose enters the bloodstream after digestion and is normally converted to glucose and so metabolised as an energy source. Sadly one infant in 40 000 lacks the necessary enzyme (known as galactose transferase or GALT) to make this conversion. They experience high accumulations of galactose and become very sick if the disease is not diagnosed. Happily for these infants a lactose-free diet is a complete remedy.

The parents of such infants are always found to be normal, possessing the missing enzyme. In families where this occurs there are more often than not other normal children. The galactose transferase (GALT) enzyme can be detected in the blood of all normal people by a **qualitative test** that shows whether it is present or not. But a **quantitative test** may also be made. In such cases it has been found that the parents of galactosaemic children, although unaffected, have *half* as much GALT enzyme as normal people. They also seem to have a one in four chance of producing children with the disease. The Mendelian explanation of this condition should now be apparent. If the allele for making the GALT enzyme is represented by the letter G the recessive allele that fails to work may be represented as g. In the case described above, heterozygous parents of genotype Gg are phenotypically normal. Although they have only one functional allele it seems to produce enough enzyme for their complete health. As these parents will produce half of their gametes with a defective allele, the chances are that one quarter of their offspring will be homozygous gg. (If you do not follow this go back to section 4.4.)

A geneticist might describe this pattern of inheritance as **completely dominant**, for the heterozygote has no outward phenotypic sign of deficiency. However, biochemically the heterozygotes have only one allele acting in each cell and not two; only half as much enzyme is produced. In this sense the phenotype is intermediate and the allele could be described as showing an **incomplete dominance**.

Pink snapdragons and roan cattle

Classical examples of such **incomplete dominance** can be demonstrated in all sorts of plants and animals. In the snapdragon, *Antirrhinum*, the cross of red-flowered plants with white-flowered plants gives pink. The pink flower colour is due to a single functional colour allele providing half as much red pigment as normal. Two pink flowers crossed together produce a red, pink, and white offspring in a ratio of 1:2:1. This is a typical ratio for an F.2 in which incomplete dominance is involved.

One of the clearest examples of incomplete dominance in mammals is that of coat colour in the breeds of shorthorn cattle and in some horse coat colours. Here a cross between red-brown-coated animals and white-coated animals produces an intermediate F.1 coat colour described as roan. Close examination of the roan coat reveals it to consist of hairs that are either red-brown or white, the two parental colours. It would seem that either one or the other of the alleles is switched on in the formative group of cells that produce each hair.

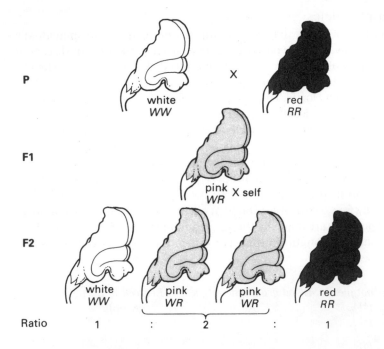

P white *WW* X red *RR*

F1 pink *WR* X self

F2 white *WW* · pink *WR* · pink *WR* · red *RR*

Ratio   1   :   2   :   1

Fig. 4.5   Incomplete dominance in snapdragons

Codominance: equal expression of different functioning alleles

In the examples given so far of incomplete dominance the recessive allele seems to be non-functional. GALT-minus allele (*g*) cannot produce the enzyme; the 'white' alleles of cattle coat colour or snapdragon flowers are pigment deficiencies, the absence of which could be attributed to missing enzymes. But there is another type of allelic partnership in which both allelic sets of DNA certainly do function, each coding for a different expressed phenotype. Neither is recessive so they are termed **codominant**.

Codominance is well illustrated by many of the **antigenic blood proteins**. Two such 'native' antigens, proteins M and N, are produced on the surface of red blood cells. The gene locus that they come from is called 'L' (after Landsteiner, the discoverer of blood groups) and is found on human chromosome 4. There are three possible genotypes and phenotypes for this blood group.

Genotype	Phenotype	Native antigens on RBC
$L^M L^M$	M	M only
$L^M L^N$	MN	M and N
$L^N L^N$	N	N only

In this case we clearly have *two* functional alleles that are both transcribed and translated into protein in the *same* red blood cell. Both alleles are switched on together.

41

4.6   Multiple alleles

Although only two alleles may be present at one locus in an individual's genotype, there may of course be more than two allele varieties in the breeding population as a whole. The existence of **multiple alleles** is well established in experimental animals such as *Drosophila* and mice, where dozens of alleles may be involved at one gene locus. The simplest and first discovered example of this in human genetics is the **isoagglutinogen** locus 'I', on human chromosome 9, that governs the most well known of blood groups, the **ABO system**. Here there are three alleles, any two of which may be in the genotype of an individual. I^A codes for an enzyme synthesising an antigen A on the surface of red blood cells. I^B codes for the production of antigen B. These two are codominant and either of them on its own is dominant to the third possible allele i which is recessive, coding for no antigenic protein at all. In its homozygous condition, ii, no antigenic proteins are made by the gene; amongst the different ABO blood groups 'O' stands for nought. In the ABO system there are six genotypes and four phenotypes.

Genotype	Blood group phenotype	Red cell antigen
$I^A I^A$	A	A only
$I^A i$		
$I^B I^B$	B	B only
$I^B i$		
$I^A I^B$	AB	A and B
$i\ i$	O	neither

The correct identification of blood groups is an essential first step in successful blood transfusion from donor to recipient. Antigens A and B are cell markers which, amongst many other blood cell markers, give that particular individual their blood tissue type. All tissues have antigenic cell surface proteins. These help our bodies to recognise their own cells and to attack the cells that are foreign to the body. The largest array of multiple alleles known in humans is the **HLA** complex, on chromosome 6. Here there are four gene loci each of which may have from five to thirty different alleles! HLA (Human Lymphocyte A) antigens are proteins on the surface of the body's defence cell system. An individuals' 'tissue type' may be determined by testing their white cells for these proteins. This procedure is the basis of telling whether tissues are suitable for organ transplantation. A suitable donor is one in which the numerous HLA antigens and their multiple alleles are the same as those in the recipient. Because of the genetic basis of the HLA system, close relatives are the most likely to be suitable donors. Identical twins may exchange organs with almost no problem of tissue rejection.

4.7   Independent assortment in human genetics

When Mendel experimented with the inheritance of more than one pair of contrasting characters he discovered that the pairs of characters behaved independently. As we are now able to see, either of one pair of alleles on one pair of chromosomes may join in a gamete with either of another pair of alleles, with equal probability. Referring back to the example of Mendel's dihybrid pea cross (p.32) where the F.1 had a genotype of $AaBb$, the allele A might go towards the formation of a gamete equally with either allele B or b. Again with equal chance, allele a might form a gamete with either B or b. This led us to see four possible gamete genotypes (AB, Ab, aB, ab,) each separately able to combine with each of the same four to produce 16 combinations in a phenotypic ratio of 9:3:3:1.

Such neat ratios are only obtainable with very large numbers of offspring; such is not possible in human family sizes. What is possible, however, is the accurate prediction of ratios on the basis of Mendelian theory. This is set out next and we shall later apply it in predicting the probability of the occurrence of certain human single-gene defects.

Let us consider a cross between a person of blood groups M and O and a person with blood groups N and AB. Our knowledge that the MN system gene 'L' is on chromosome 4 and that the ABO system gene 'I' is on chromosome 9 enables us to predict that as each chromosome pair will segregate independently, either of the two alleles at one locus may go with either of the two alleles at the other locus in gamete formation. Further, we may then predict the genotypes and the phenotype ratios that would occur if large numbers of offspring were produced. (See Fig. 4.6) This exercise shows that half the offspring will be group A and half group B and all of group MN. This cross should be quite easily followed.

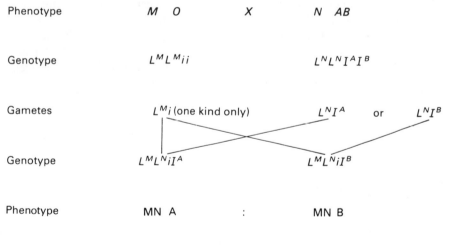

Phenotype	$M \quad O$	X	$N \quad AB$
Genotype	$L^M L^M ii$		$L^N L^N I^A I^B$
Gametes	$L^M i$ (one kind only)		$L^N I^A$ or $L^N I^B$
Genotype	$L^M L^N i I^A$		$L^M L^N i I^B$
Phenotype	MN A	:	MN B
Ratio	1	:	1

Let us take a much more complex example employing the same genetic diagram method. If two people have the same blood group phenotype MN AB, how many offspring phenotypes may there be and in what proportions? This time the answer is more complex; for each of the crosses, MN X MN and AB X AB would on their own produce a 1:2:1 ratio. Here, therefore, a 1:2:1 ratio for one set of alleles overlies another 1:2:1 ratio for the other. (See Fig. 4.7)

Phenotype   MN   AB   X   MN   AB

Genotype $L^M L^N$ $I^A I^B$ $L^M L^N$ $I^A I^B$

Gametes (for both) $L^M I^A$, $L^M I^B$, $L^N I^A$ $L^N I^B$

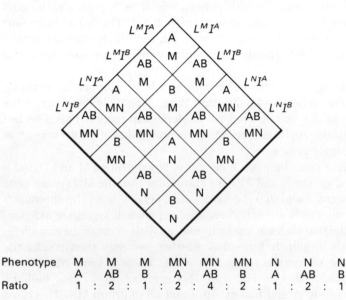

Phenotype	M A	M AB	M B	MN A	MN AB	MN B	N A	N AB	N B
Ratio	1	2	1	2	4	2	1	2	1

The elegance of Mendelian genetics, so far as human genetics is concerned, lies in its predictive value. One can, taking the example above, predict that two parents of MN AB phenotype have only a $\frac{4}{16}$ or one quarter chance of producing children with the same genotype and phenotype as themselves. Genetic prediction is of great importance in the calculation of disease risks, as will be seen when hereditary diseases are considered in Chapter 8.

5 Sex-determination and sex-linked inheritance

5.1 The determination of sex

There is no neater example of the segregation of chromosomes in meiosis than the separation of the **sex chromosomes**, X and Y, in the formation of a spermatozoon. In the testis each spermatocyte undergoes two divisions, resulting in the formation of four spermatids. Two of these bear X chromosomes and two bear Y chromosomes. Half of all the sperms produced are therefore X-bearing and half are Y-bearing. At the fertilisation of the ova, all of which carry a single X chromosome, the genetic constitution of the new individual will be reconstituted: XX for female and XY for male.

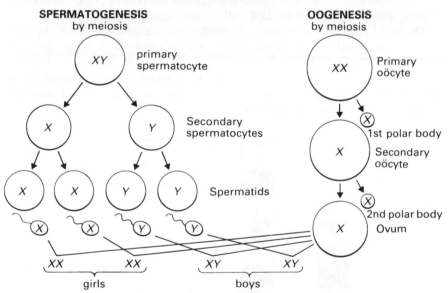

Fig. 5.1 The segregation of X and Y chromosomes in spermatogenesis determines the sex of future boys and girls.

Although equal numbers of X and Y sperms are produced, Y-bearing sperms appear to be able to achieve a higher rate of successful fertilisations. This has not been entirely proven but if this is not the case then there must be a higher rate of successful implantation of male embryos. This must be so, for a higher proportion of embryos and foetuses that are male are naturally aborted during pregnancy yet despite these losses there are still fractionally more births of boys than of girls. Some unknown mechanism is creating a pro-male bias to balance the losses. In the United Kingdom, there are 105 boys born for every 100 girls.

45

Because more boys than girls are also prone to die in childhood, the ratio finally approaches equality by the age of marriage. This at last makes the sex ratio sportingly fair, but things do not remain equal! Although a man would have the potential to leave more offspring than a woman in a polygamous society, men live less long on average. In Britain today, there is a five-year difference between the life expectancy means and by the time one reaches the 85-year-old age-class, women outnumber men by 2:1. Men may claim some advantages in life, but as we shall see in this chapter males are, with respect to genetic health, undoubtedly the weaker sex.

In different groups of organisms the mechanism of sex determination varies widely. In many insect species there is no Y chromosome at all, males having a single X chromosome and females having two. Amongst bees and ants the queen and workers are diploid and the males haploid. In the fruit fly, *Drosophila*, there is the same XX/XY system as in humans, but in the mosquito sex is determined by a single gene. In some animals such as birds and butterflies the XX (female) and XY (male) pattern is reversed. A hen is XY and a cockerel is XX. The sex of a chicken is therefore detemined by the genotype of the ovum and not the genotype of the sperm. In some reptiles (e.g. *Crocodilus*) the determination of sex is not even genetically based but is determined by embryological development temperature. The hotter crocodile eggs in the nest (> 34°C) all become males, whilst cooler eggs (< 31°C) all become females.

Of the 23 pairs of chromosomes in the human genome the two X chromosomes of the female and the X and Y chromosomes of the male merit the name sex chromosomes. Not only do they govern the different pathways of sexual development but the genes they bear often show an inheritance pattern related to the sex of the individual. Genes on the sex chromosomes are responsible for what is called **sex-linked inheritance**. (See Fig. 5.2)

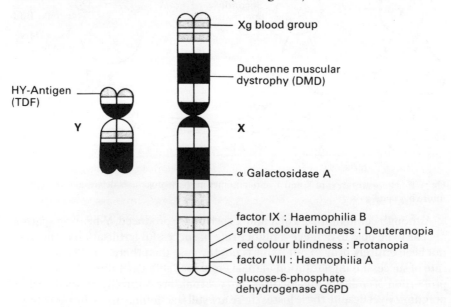

Fig. 5.2   The human sex chromosomes showing the locus of several genes.

The X chromosome is much larger than the Y. The two differ in shape and banding pattern and in the genes that they carry. They have sufficient homologous loci close to the centromere for the chromosomes to recognise each other and pair up at meiosis in the male. The 22 other pairs of chromosomes in the human genome are not related to sex determination and are termed **autosomes**. Inheritance is therefore for the most part **autosomal inheritance**. Autosomal genes may affect the phenotype of individuals of one sex more than individuals of the other (see 5.3), but they play no part in steering the pathway to maleness or femaleness in development.

5.2   What makes a male or a female?

In answer to the question 'why do human males and females develop differently?' the answer must lie either in the fact that females have two X chromosomes, not one, or that males possess a Y chromosome that females never possess. For a woman to develop normally, certainly more than one X chromosome is needed, but if a functional Y chromosome is present the distinctive male genetical characteristics *always* develop. An understanding of the sex-directing effects of these chromosomes is important in interpreting the physical handicaps that result if an individual does not have the normal sex chromosome constitution. (See also 7.6)

Early in the life of an XY bearing embryo (future male) a protein called the **HY-antigen** is produced in cells decoding a gene on the Y chromosome. This antigen, acting as a **testis determining factor**, changes the undifferentiated embryonic gonad into a testis. If there is no Y chromosome, and so no antigen, development follows a female line. In this case the gonad forms an ovary, and partly under the maternal hormone influence female reproductive structures are developed. Embryonic males, by contrast, develop a testis which produces male hormones from a very early embryonic age. The hormones inhibit the development of the female ducts and promote the growth of penis, testes and scrotum. The Y chromosome gene that triggers off the HY-antigen is the key factor in selecting the path of sexual development. The sex of an early foetus may be discovered indirectly by chromosome analysis whilst fetoscopy makes direct observation of its sex possible. (See also 8.6)

5.3   Sex-linked inheritance

Genes on the sex chromosomes are said to be sex-linked. Sex-linked patterns of inheritance differ from autosomal inheritance because the phenotypic expression of the sex-linked genes follows the chromosomes as they are passed down through a family's pedigree. Whereas X chromosomes are passed to both sons and daughters, Y chromosomes are only passed from father to son.

Y-linkage

Although men have many male features that are absent from women, most of these are the secondary result of the action of male sex hormones, coded for by autosomal genes, operating on phenotypic expression during development. Beards for example develop in response to male sex hormones and may be possessed by men and women. Both sexes produce small amounts of the opposite sex's hormones, i.e. a trace of oestrogens in males and a trace of androgens in females. Masculinisation and feminisation are thus more likely to be due to hormonal disorders of the steroid secreting tissues than they are to direct chromosomal differences. All this means that Y-linkage is extremely hard to prove. The gene for the control of the HY-antigen is the one certain example (see 5.2). Famous text book illustrations of hairy ears in men (hypertrichosis of the pinnae) are now *not* believed to illustrate Y-linkage (Jenkins 1983).

X-linkage

The vast majority of sex-linked genes that are known are X-linked. Fig. 5.2 gives the locus for several of these. In a woman (XX), the pair of X chromosomes seem to behave as a normal pair of homologues; the genes are represented by paired alleles and crossing over may occur in meiosis. It will be seen from Fig. 5.2 that three classical X-linked traits are Duchenne muscular dystrophy (DMD), various forms of colour-blindness and haemophilia A. At each one of these different loci there is normally a functioning allele in a male (XY) so that such a person would not suffer from muscular dystrophy, colour blindness or haemophilia. However, at each of these gene loci there are known to be recessive alleles that may, in the absence of a functioning partner allele, express

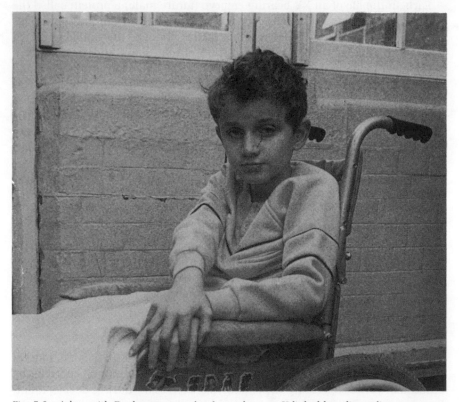

Fig. 5.3   A boy with Duchenne muscular dystrophy – an X-linked hereditary disease

the genetic diseases. For men (XY), who possess one X chromosome, the recessive conditions express themselves much more frequently than in women, where expression will only occur if they are homozygous recessive for *both* alleles. A condition like colour-blindness is of minor importance, but some X-linked conditions are near lethal in their effects and a cause of much illness and distress. (See the clinical notes on the diseases Duchenne muscular dystrophy and Haemophilia A, appended to Chapter 7.)

Colour-blindness in the U.K. is found in some form in approximately one male in every twelve. This proportion reflects the proportion of affected chromosomes; one X chromosome out of every twelve carries a recessive colour-blindness allele. There are many kinds of colour-blindness and there is certainly more than one locus involved. (See Fig. 5.2.) Approximately 1.2% of men (1 in 80) suffer from red-blindness, protanopia, and are quite unable to tell this colour from green. As woman have two X chromosomes their chances of having two recessive alleles are much reduced ($\frac{1}{80} \times \frac{1}{80}$). Such conditions therefore occur in women at a frequency that is the square of the frequency that it occurs in men. This is why only one woman in about 6400 suffers from protanopia. The masking effect of the partner dominant allele means that although women rarely ever express such conditions, they may very often be the carriers and thus able to pass on the deficient allele to both their sons and daughters. Because daughters receive X chromosomes from their father and from their mother they will express recessive conditions much less often than will sons. Sons are therefore more at risk from X-linked recessive hereditary diseases. Whenever such diseases occur they are attributed to the X chromosome which must be maternal in origin.

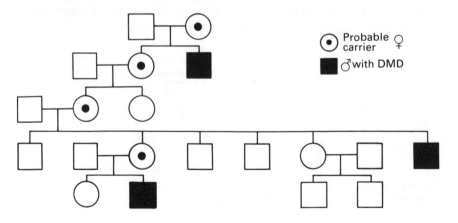

Fig. 5.4 A family pedigree showing the inheritance pattern of an X-linked recessive allele, Duchenne muscular dystrophy.

It is useful to present the chromosomes as bars to show the sex-linked patterns of inheritance. The gene locus may be marked on the bar to show its position. Fig. 5.5 sets out the typical pattern of inheritance of an X-linked recessive character such as colour-blindness. The pattern of transmission is from a mother to half her sons, with half her daughters being carriers. If the mother is homozygous for the recessive allele and therefore also expressing the phenotype, then all her sons will express the condition too and all of the daughters will be carriers.

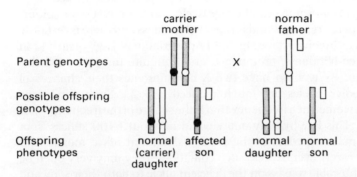

Fig. 5.5   The pattern of X-linked inheritance 1. Maternal chromosomes are shaded. Each of the offspring receive one of the two parental sex chromosomes in four possible combinations. A recessive allele (solid) may be passed by a mother to both sons and daughters, but it will only be expressed in the sons, providing that the father is normal.

If the father expresses a recessive genetic condition that is X-linked clearly this will be passed to all daughters. It is therefore the daughters of colour-blind fathers that pass on the condition and not the sons. (See Fig. 5.6.)

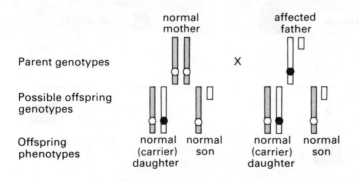

Fig. 5.6   The pattern of X-linked inheritance 2. Maternal chromosomes are shaded. The recessive allele, expressed in the father, is passed to all of the daughters. Sons are always unaffected, provided that the mother is not herself a carrier. Were the mother to be a carrier, in this case, half of the daughters would be homozygous for the recessive allele and therefore affected.

The pedigree pattern of X-linked transmission is shown in Fig. 5.4. Just such a pattern as this was shown in the pedigree of the British Royal Family in the nineteenth century where Queen Victoria, a carrier of haemophilia A, passed on the allele via her children to three other royal households. Queen Victoria must have received a mutant allele from one of her parents.

Fig. 5.7   Queen Victoria's family. Queen Victoria (seated) with her husband, prince Albert (left) and their nine children (from the left: Alfred, Helena, Alice, Arthur, Beatrice (baby), Victoria, Louise, Leopold and Edward. Alice and Beatrice, at least, were carriers of haemophilia A, whilst Leopold was afflicted.

5.4   Gene dosage: the Lyon hypothesis and tortoise-shell cats

One of the imbalances that results from females having two X chromosomes and males only one, is that females will receive a double allele dose compared to a single one in males. (The presence of an extra chromosome or the absence of a chromosome may be traumatic to a person's well-being, as will be apparent when such conditions as Down's Syndrome and Turner's Syndrome are considered in Chapter 7.) As males must get enough genetic information from one X, one might expect females to make some compensation for their double dose. They appear to do this by effectively locking up one of the two X chromosomes from about the 2000 cell stage of the embryo. A female is thus a mosaic of the two X chromosomes, either one or the other being inactivated. This was first suggested by Mary Lyon in 1961 and is referred to as the **Lyon hypothesis**. The X-inactivation is microscopically demonstrable in cheek epithelial cells and white blood cells where a darker chromatin patch, or **Barr body**, reveals the position of the condensed inactive chromosome. The Lyonisation of all but one X chromosome is dramatically illustrated by the presence of *two* Barr bodies in the nuclei of cells taken from women with an abnormal set of *three* X chromosomes. The presence of Barr bodies enables forensic scientists to distinguish male from female blood.

The most dramatic display of X-inactivation is in the female tortoise-shell cat. Coat colour in domestic cats is governed by many different gene loci. One of these, the 'O' locus, is X-linked. Here there are two possible alleles O for orange (ginger or marmalade) and the other + for tabby or black. (The actual non-orange coat colour is governed by autosomal coat colour genes at a different locus.) As males have only one X chromosome they can only be entirely orange or entirely non-orange (tabby/black). Male genotypes are therefore O or +, whilst female genotypes are of three types, OO, O+ or ++. Females, however, with two X chromosomes, will have one of the two Lyonised at random from an early embryonic stage. Thus patches of the coat express the allele present on one of the X chromosomes whilst other patches express the other. If the female cat is homozygous at the 'O' locus it will not be a tortoise-shell, but heterozygosity will give the tortoise-shell cat its characteristic blotchy appearance. The gene frequency of the ginger allele in Britain is about 0.17. This means that approximately 28% of female cats will be tortoise-shells.

Lyonisation certainly occurs in women. The enzyme glucose-6-phosphate dehydrogenase (G6PD), involved in respiration in blood cells, is X-linked (see Fig. 5.2). In women that are heterozygous at this locus, approximately half their red blood cells possess the enzyme and a half do not. We do not know if such mechanisms to reduce the gene dose apply throughout the body. Lyonisation seems not to influence colour-blindness in women where it might be expected to increase it, indeed it is sometimes claimed that women see colours much more vividly than men. In the case of haemophilia, female carriers seem to produce enough (A.H.G.) factor VIII to protect themselves fully from the disease. Sex-linked inheritance is important, not least because it shows that genetically all is not equal between the sexes.

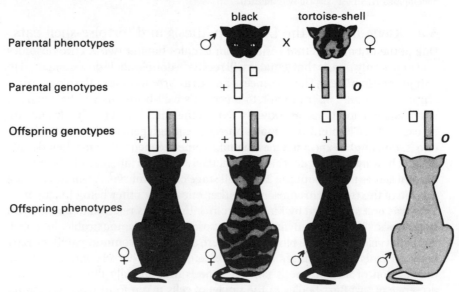

Fig. 5.8 The inheritance of coat colour in cats. The gene governing orange and normal colouration is on the X chromosome. Cats having both the orange allele and the normal colour allele as a partner will be tortoiseshell and female.

6 Complex patterns of heredity

6.1 Introduction

Genetics has grown up as a very experimental science; carefully controlled crosses of animals and plants often produce neat ratios in subsequent generations enabling us to see how chromosomes have segregated at meiosis and come together again in pairs in a new individual. But many patterns of inheritance are not so easy to interpret, for members of different pairs of alleles may be tied together on the same chromosome, they may interact with each other, have more than one effect upon the phenotype or sum together with one another in their influence. Alleles may be influenced by the hormonal environment of the body in the way that they are expressed. They may even so bias the survival of their bearers that expected ratios and numbers of offspring are altered. This chapter sets out to explain these more complex patterns of human inheritance; these are respectively: **autosomal linkage, epistasis, pleiotropy, polygenic inheritance, sex-limited inheritance** and **lethal genes.** Because human heredity is not subject to experiment it is often difficult to give clear examples of these patterns from human genetics alone. In such cases better known animal and plant illustrations are provided.

6.2 Autosomal linkage

Early experimental studies of the genetics of maize revealed that sometimes Mendel's second Law of Independent Assortment (that either of one pair of alleles would freely combine with either of another pair in a single gamete) did not apply. Where two such pairs of characters were chosen and the plants crossed together to produce an F.1 hybrid, the selfed F.1 did not produce a 9:3:3:1 ratio in the F.2 but would produce a greater proportion of the two parental types. The parental contributions seemed not to have been free to mingle and separate as much as Mendel had predicted. This linkage together of parental types was soon perceived as being due to two pairs of alleles being on the *same* pair of chromosomes; they were literally tied together.

In one example maize was investigated for the inheritance pattern of cereal grain colour and shape. Most maize inheritance patterns are straightforwardly predictable in a Mendelian manner, but some show this phenomenon of linkage. If a pure-breeding colourless and shrunken-seeded plant (*ccff*) is crossed with a pure-breeding coloured and full-seeded plant (*CCFF*) the F.1 is all coloured and full-seeded (*CcFf*). As will be seen from the allele notation taken, coloured and full were identified as being dominant to colourless and shrunken. If this hybrid (*CcFf*) was crossed back to the doubly homozygous recessive parent (*ccff*) a 1:1:1:1 offspring ratio might be expected, reflecting the four possible gamete genotypes produced by independent assortment. On account of this linkage phenomenon it does not occur. Fig. 6.1 sets out the expected and actual result.

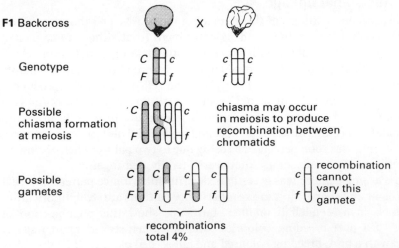

P	parent seed phenotype			colourless shrunken	coloured full
	Parent plant genotype			*ccff*	*CCFF*

| F1 backcross | Seed phenotype | | | | *CcFf* X *ccff* |

| | Gametes | *CF* | *Cf* | *cF* | *cf* | | *cf* |

		coloured full	coloured shrunken	colourless full	colourless shrunken
Expected ratio		1 :	1 :	1 :	1
Expected %		25	25	25	25
Actual %		48	2	2	48

Fig. 6.1 Maize back-cross revealing linkage. Dihybrid F.1 backcrossed to the double homozygous recessive fails to give the expected 1:1:1:1 ratio.

The explanation of this result is not hard to find if we assume that the two pairs of alleles are on the same chromosome pair but at two different loci. (See Fig. 6.2.)

F1 Backcross

Genotype

Possible chiasma formation at meiosis — chiasma may occur in meiosis to produce recombination between chromatids

Possible gametes — recombination cannot vary this gamete

recombinations total 4%

Fig. 6.2 The non-independence or linkage of characters is explained by their physical linkage on the same chromosome. Recombinations occur when chiasmata separate linked alleles.

Clearly the *CF* linkage and the *cf* linkage will make the original parental phenotypes occur more frequently. The problem is not to see why there are more of these, but how the less frequent recombinations occur. The answer to this lies in unlinking the linked genes at meiosis by a cross-over. This is shown as it might occur in Fig. 6.2. This makes possible the four offspring phenotypes that we would normally expect to occur in independent assortment. However, if a chaisma were to take place every time between a pair of alleles, as is shown here, then of course a 1:1:1:1 ratio would follow, but the chances of such a cross-over happening at every meiosis between two physically close gene loci is small.

The discovery by geneticists, such as T. H. Morgan, that the linkage phenomenon occurred, led them to think of the loci of the genes as being physically linked on the same chromosome. Much to their delight they discovered that the number of linkage groups that were to be found in species was the same as the number of pairs of chromosomes. Thus maize had ten linkage groups, the pea had seven and the fruit fly *Drosophila* had just four. These different species had, respectively, ten, seven and four pairs of homologous chromosomes. This discovery was only the beginning of what was to become an ingenious development – the mapping of genes on chromosomes.

It became clear to the early geneticists that where the genes were linked the frequency of recombination was a direct measure of how close together the gene loci were. Thus if two loci were very close together then a cross-over between them would occur more rarely, but if they were far apart separation was much more likely to happen. (See Fig. 6.3.)

Look back to the example of linkage in maize. If cross-overs had occurred every single time between the two loci, the percentage recombination would be 50%. In this particular case the percentage recombination was 4%. Thus crossing over between the loci was not common. We would describe the linkage as 'strong' and would predict that the gene loci were fairly close together.

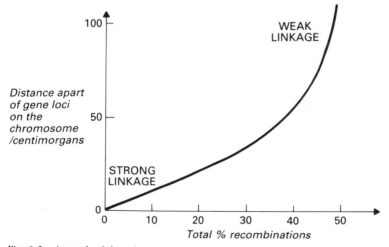

Fig. 6.3   A graph of the relationship between percentage recombination and the spatial separation of gene loci. The lower part of the graph is linear. On long chromosomes alleles at distant loci assort independently in a Mendelian manner.

6.3 Linkage in humans

Linkage as a phenomenon in human inheritance has a lumping effect; groups of genes that might otherwise behave independently of each other are inherited in association together. Because all the alleles from our parents are not completely scrambled up, groups of features may be inherited in a block. Only the very closest of genetical studies can find linkage groups, but it is just such an exercise that is needed if we are to discover where the genes that cause genetic disease are located. It was such an exhaustive exercise that made possible the mapping of the human X chromosome (Fig. 5.2 in Chapter 5). It will be seen that the colour-blindness gene and the haemophilia gene loci are very close together. In families where a mother produces some sons that are colour-blind and other sons that are haemophiliac, they very rarely produce sons that are both. Conversely, in families where a mother produces individual sons that are *both* colour blind and haemophiliac, it is very common for other sons to have neither condition. This suggests to the medical geneticist that the loci are close together; where the alleles are on different X chromosomes, in the mother, they rarely cross-over and so tend to be inherited in a group together. This is certainly not the case with colour-blindness and Duchenne muscular dystrophy; these are far apart on the X chromosome.

Human geneticists are on the look out for such **linkage patterns**. PTC tasting, for example, is linked to the Kell blood group on chromosome 7. Fig. 6.4 shows the inheritance of a dominant genetic handicap, nail-patella syndrome, in a family pedigree. The MN and ABO blood groups of the individuals are also recorded. As was seen in Chapter 4, the MN and ABO blood groups are inherited independently of each other in a Mendelian manner. They follow Mendel's Law of Independent Assortment, for the MN locus is on chromosome 4 and the ABO locus is on chromosome 9. Study the pedigree diagram and decide to which blood group system the nail-patella syndrome is linked. On which chromosome is the defective allele likely to be? Can you explain why the probability of this third generation occurring by random assortment is a 1 in 128 (i.e. $(\frac{1}{2})^7$) chance?

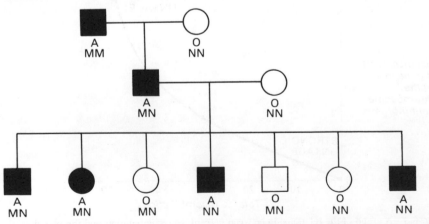

Fig. 6.4 The pedigree of a family with nail-patella syndrome. This genetic condition is linked to one of the blood group loci (see text).

6.4   Epistasis and pleiotropy

Epistasis and pleiotropy are two common types of inheritance in humans.

Epistasis

Epistasis is the masking by one gene of the effects of another gene (i.e. one at a different locus); as such it is a form of **gene interaction**. In mice the allele for black coat colour *b* is recessive to the normal brown fur allele *B* at the 'B' locus whilst the gene governing the production of pigment in the fur is at another locus 'C', on a different pair of chromosomes. If a mouse is homozygous recessive for the pigment deficiency allele at this second locus it will be an all white (albino) mouse, of genotype *cc*. The dominant colour allele *C* must be present in the genotype, as *CC* or *Cc*, if the phenotype of brown or black is to be expressed.

In a cross between a doubly homozygous black mouse and an albino mouse, homozygous at the brown locus, the F1 will be brown but the F.2 will be brown to black to albino in a ratio of 9:3:4.

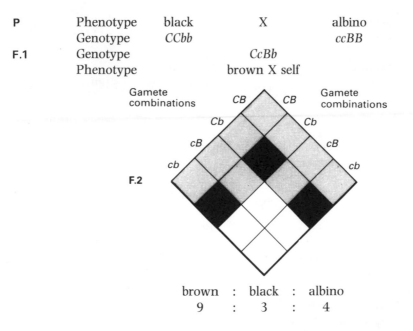

P	Phenotype	black	X	albino
	Genotype	*CCbb*		*ccBB*
F.1	Genotype		*CcBb*	
	Phenotype		brown X self	

F.2

Gamete combinations   CB CB   Gamete combinations
                          Cb   Cb
                          cB   cB
                          cb   cb

brown : black : albino
  9   :   3   :   4

This ratio is most easily seen to be a 9:3:3:1 ratio that has been modified by the epistatic effect of the albino allele. One way of viewing such gene interaction is to say that more than one gene locus affects the *same* phenotype.

Gene 1
        ⟩Phenotype
Gene 2

An example of two different genes that affect the same phenotype is illustrated in humans by albinism. Most often albinos are the children of two people that do not show the condition. In such families it occurs with a 25% chance, indicating a recessive condition at one gene locus.

Phenotype	normal	X	normal
Genotype	Aa		Aa

Offspring		albino	($\frac{1}{4}$ chance)
		aa	

Albinism is caused by one enzyme being defective in a reaction sequence that leads to the skin and hair pigment melanin. If however there are two 'A' loci for albinism 'A^1' and 'A^2', that code for two different enzymes, albinos may be of two different genotypes, a^1a^1 or a^2a^2. Both loci may contribute to a lack of hair and skin pigment. Because the two different genes have the same effect they are termed **mimic genes**. This is the explanation for the fact that it is quite possible for two albinos to produce normally pigmented children.

Phenotype	Albino	X	Albino
Genotype	$A^1A^1a^2a^2$		$a^1a^1A^2A^2$

Offspring	Normal pigment	
	$A^1a^1A^2a^2$	(all)

Pleiotropy

Pleiotropy contrasts with epistasis, for here one gene locus finds expression in more than one phenotype.

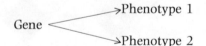

Pleiotropy is encountered by students investigating the inheritance pattern of the vestigial wing gene in the fruit fly *Drosophila*. Not only does this allele cause a reduction of the wings to little stubs, but it also modifies the fly's balancing organs, distorts some of its bristles, alters the shape of the reproductive organs, lowers its egg production, and hence its potential to reproduce. A single mutant allele can have similar multiple effects in human beings. In the genetic disease **cystic fibrosis** the recessive allele causes a much thicker than normal mucus to be secreted. This has the pleiotropic effect of not only causing lung congestion but also pancreatic duct obstruction, which may lead in turn to digestive difficulty and also to pancreatic damage, tragically leading on to diabetes. (See the descriptive note on cystic fibrosis in Chapter 7.) It is quite interesting to speculate upon and indeed investigate possible pleiotropies that are not harmful. Is the possession of red hair and freckling of the skin an example of pleiotropy? Are there some hair colour genes that also influence eye colour? What do the data in Table 6.1 suggest?

Table 6.1 The relationship between different eye colours and different hair colours amongst Aberdeen school children

Eyes	Hair					
	red	fair	medium	dark	black	Totals
blue	131	1105	885	348	1	2470
light brown	405	2285	2434	851	9	5984
medium brown	360	1208	3242	1601	29	6440
dark brown	209	366	1621	2094	95	4385
Totals	1105	4964	8182	4894	134	19 279

Source: J. F. Tocher, Biometrika 6

6.5 Polygenic inheritance

Early geneticists found that many patterns of inheritance in different organisms fitted into the Mendelian picture of simple ratios with clear dominance and recessiveness. The phenotypes could be classified into discrete variables, that is to say separate classes in a **discontinuous variation**. Thus for example in peas there was a tall and a dwarf variety and crosses between the hybrids produced not a range of heights but discrete classes of tall ones or short ones. More commonly, in other genetic experiments with both plants and animals, such a situation did not seem to apply and a more evenly spread variation over the range of phenotypes was observed. Such cases, without discrete class groupings, were therefore described as displaying **continuous variation**.

In 1909 Wilhelm Johannsen made an important first step in showing that continuous variations, which in humans include such things as height, weight and skin colouration, had two component influences. These were **heredity** and **environment**. It is widely recognised that being tall or short or fat or thin is not just a question of genes or of diet! (The discussion of 'nature and nurture', the relative influence of heredity and environment in humans, is tackled in Chapter 9.) Johannsen, who studied plants not people, ingeniously devised a way of discriminating between these influences. If you have ever taken beans out of a pod you will know that their size is not equal. Are the large ones big because they had the best place in the pod? Or, on the other hand, is there a real genetic difference that affects their size? Johannsen took a freely interbreeding variety of *Phaseolus vulgaris*, the dwarf bean, and selected from the initial variety the heaviest and lightest beans and then proceeded to breed two pure lines from each stock, by self-pollinating within the lines to produce as much genetic uniformity as possible. Not surprisingly, he discovered that the line he developed from the smallest beans was on average smaller than the line that he developed from the largest beans, when grown under the same conditions. However there was still a small spread of size in each line even when they were pure-breeding and virtually homozygous at every gene locus. Although every bean in each strain was genetically the same there was still a lot of difference in weight depending on how well each plant, pod and seed had grown. By contrast, he developed several distinct pure-breeding lines over a wide range of bean sizes. Here the consistent *mean* differences must have been due to many different inherited factor's governing the one variation. This is now temed **polygenic inheritance**. Johannsen's experiment clarified our thinking by showing that beans vary in size (phenotype) because of the range of their inherited differences (genotype) *and* the quality of their conditions of growth (environment).

| 1 | 0 | 0 | 1 | 5 | 7 | 7 | 23 | 25 | 26 | 27 | 17 | 11 | 17 | 4 | 4 | 1 |
| 4'10" | 4'11" | 5'0 | 5'1" | 5'2" | 5'3" | 5'4" | 5'5" | 5'6" | 5'7" | 5'8" | 5'9" | 5'10" | 5'11" | 6'0 | 6'1" | 6'2" |

Fig. 6.5 Army recruits lined up according to their heights. Such continuous variation, like that found in Johannsen's beans, is a product of both polygenic inheritance and environmental factors in growth.

At the same time (1909) that Johannsen was making clear that the phenotype reflected an array of influences, studies of the red kernel-colour of wheat quantified the genetic component. In wheat, *Triticum vulgare*, there are three independent loci, each possessing alleles that give an additive red effect to the seed. Thus $R^1R^1R^2R^2R^3R^3$ is the genotype of the dark red fully homozygous form and $r^1r^1r^2r^2r^3r^3$ is the genotype of the white triple homozygote. A cross between the darkest red and the white phenotype gives an intermediate red $R^1r^1R^2r^2R^3r^3$. Much as the snap dragon F.1 hybrid shows segregation to produce red : pink : white in a ratio of 1:2:1, so the intermediate red wheat F.1 plant shows segregation of *each* of the three pairs of alleles to produce a range of phenotypes from dark red at one extreme to white at the other. Each additional R allele has an additive effect. (See Fig. 6.6.)

The resulting F.2 ratio is 1 : 6 : 15 : 20 : 15 : 6 : 1, ranging from dark red through intermediate red shades to light red and white.

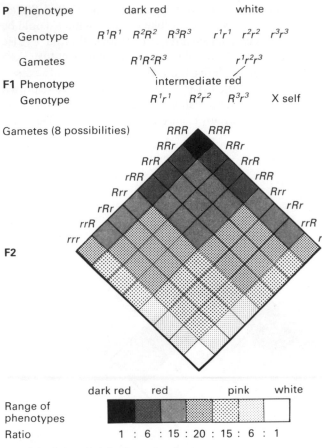

P Phenotype	dark red			white		
Genotype	R^1R^1	R^2R^2	R^3R^3	r^1r^1	r^2r^2	r^3r^3
Gametes	$R^1R^2R^3$			$r^1r^2r^3$		

F1 Phenotype intermediate red

Genotype R^1r^1 R^2r^2 R^3r^3 X self

Gametes (8 possibilities)

F2

Range of phenotypes: dark red red pink white

Ratio 1 : 6 : 15 : 20 : 15 : 6 : 1

Fig. 6.6 Polygenic inheritance in wheat.

For a pair of alleles at one locus where gene expression is additive a 1:2:1 ratio is obtained. If there were two loci acting in a polygenically additive manner there would be four types of gamete producing 5 phenotypic classes in a ratio of 1:4:6:4:1. Those familiar with Pascal's triangle, in doing mathematics, will recognise these ratios as alternate layers in its construction.

Numbers of independent gene loci											Fraction of F.2 like either original parent	Number of phenotypic classes
					1							
				1		1						
1				1	2	1					$\frac{1}{4}$	3
			1	3	3	1						
2			1	4	6	4	1				$\frac{1}{16}$	5
		1	5	10	10	5	1					
3		1	6	15	20	15	6	1			$\frac{1}{64}$	7
	1	7	21	35	35	21	7	1				
4	1	8	28	56	70	56	28	8	1		$\frac{1}{256}$	9

Fig. 6.7 Polygenic crosses and Pascal's triangle

As long ago as 1911 Davenport, in a very race-conscious United States, suggested that skin colour differences between 'blacks' and 'whites' could be explained by a polygenic system of just two independent loci, 'P¹' and 'P²'. He suggested that each dominant allele P would have an additive effect to darken the skin. In a cross between someone of pure West African descent and a person of pure Northern European descent the F.1 is certainly an intermediate brown as he suggested. In a cross between two such F.1 individuals the offspring should display a range of tones of brown from white to black. (See Fig. 6.8.)

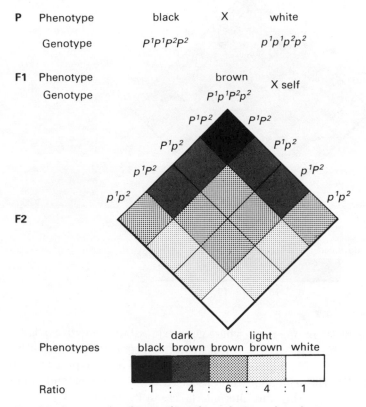

Fig. 6.8 Davenport's polygenic skin colour inheritance hypothesis.

Certainly marriages between two individuals of intermediate genotype do produce a greater variety of shades of brown but $\frac{1}{16}$ are not as white, nor $\frac{1}{16}$ as black as their original grandparents, as Davenport would suggest. Studies of many mating combinations of this kind show that it is only about 1 in 256 that would return to the extremes of the grandparental type. Using Pascal's triangle this suggests that there are *four* pairs of alleles involved and possibly more. Anyone who looks critically at 'whites' will see that they vary considerably in their pigment expression and ability to tan. Similarly indigenous African people vary over a very wide range in the number of pigment alleles that are possessed. As is made clear in Chapter 10, on the genetics of race, terms like 'black' and 'white' have very little meaning in genetics without careful definition.

A much simpler example of polygenic inheritance is that of human eye colour. It is common for students to be introduced to genetics by being told that eye colour is due to a single gene locus in which 'brown' is dominant to 'blue'. This would imply that brown-eyed parents could have blue-eyed children, but blue-eyed parents could not have brown-eyed children. There are, however, more than a few blue-eyed parents with brown-eyed children and they at least should know that this is perfectly possible!

Blue eyes have a reflective iris that shines light back out of the eye. Brown eyes have the iris pigmented with melanin granules that absorb the light more; the more granules of pigment that there are, the darker the eye. Supposing there to be three pairs of alleles at three 'B' loci, sample genotypes might be as follows:

		Number of alleles adding brown pigment
BB BB BB	deep brown	6
BB BB Bb	dark brown	5
BB BB bb	medium brown	4
BB Bb bb	light brown	3
BB bb bb	deep blue	2
Bb bb bb	medium blue	1
bb bb bb	pale blue	0

As with skin colour, each *B* allele has an additive effect. Consider now the cross between two 'blue-eyed' parents each with a few *B* alleles.

	'deep blue' $B^1b^1\ b^2b^2\ B^3b^3$	X	'light blue' $B^1b^1\ b^2b^2\ b^3b^3$
possible gamete	$B^1b^2B^3$		$B^1b^2b^3$
probability	$\frac{1}{4}$		$\frac{1}{2}$

$$B^1B^1\ b^2b^2\ B^3\ b^3$$
'light brown'
$$\frac{1}{8}$$

The chance, therefore, of such blue-eyed parents having a brown-eyed child occurs once in eight; it need have nothing to do with a brown-eyed stranger!

6.6 Sex-limited inheritance

Sex-limited inheritance is an example of a simple Mendelian inheritance pattern in which the phenotypic expression depends upon the male or female sex hormone balance in the body; it is quite *unrelated* to sex-linkage (see Chapter 5). Pate baldness is a classical example of sex-limited inheritance, due most simply to a dominant allele at a single locus. Men who bear the single dominant allele begin to lose their hair from the top of their heads by the age of eighteen. By the age of 30 they are often completely bald on top (the pate) and yet have a strong growth of beard and hair at the sides of the head. The expression of the allele seems to be directly related to the testosterone levels in the blood. Women who are homozygous for the dominant allele begin to show loss of hair in late middle age as their oestrogenic levels decline. The inheritance is in no way sex-linked but is autosomal.

6.7 Lethal genes

A lethal gene is one which directly or indirectly brings about the death of its possessors before they are able to reproduce. At first sight this notion seems ridiculous, for possessing it would seem to guarantee that it would not be inherited, but this is not so. The lethal alleles are generally fully recessive and thus may be inherited and transmitted in the heterozygous state. Such defective alleles, if paired in a homozygous state, may cause the death of the possessor before birth. This form of lethality may account for some natural abortions (miscarriages). Such an effect would bias the sex ratio in humans if it were an X-linked lethal gene. (See 5.1.) By contrast some mutant alleles are incompletely dominant, being lethal in the homozygous state but not at all life-threatening in the heterozygous condition. Two such examples referred to later are achondroplasia (dwarfism), where the dwarfed person is heterozygous, and sickle cell anaemia, where the heterozygote is near normal. The homozygotes in the first case do not survive to birth, and in the second, sickle cell anaemics have a low life expectancy.

A dominant lethal allele is not an impossibility. Huntington's disease (Huntington's chorea) is a genetic illness that is caused by a single dominant autosomal allele. This causes the death of the possessor in middle life, as a result of a degenerative nervous illness. The tragedy of this illness is compounded by the fact that patients commonly develop the symptoms of the disease only after they have had their children, to whom they may well have already passed on the disease. Fortunately in Britain the condition is extremely rare and is now being rapidly eliminated by genetic counselling. (See Chapters 7 and 8.)

Such genetic conditions can easily frighten people when they first read about them; this darker side of human genetics is examined in the next chapters. It is important to approach the problem of human genetic disease with the right attitude towards ourselves. It is astonishing that we are all so genetically healthy and it is important in appreciating this to develop our sympathy for those that have not had as good a deal as we have. Human heredity is highly complex in its patterns; we cannot know a fraction of the genetic realities we meet in our own lives, but we can accept ourselves for the complexity that we are and marvel at it.

7   Mutation: cause and effect

7.1   Introduction

A **mutation** is any heritable change in the nucleotide sequence of an organism's DNA. It is convenient to regard a chromosome as an extremely long strand of DNA replicated with quite remarkable faithfulness between cell divisions. It is not surprising that some mistakes should occur and that these be passed on to a subsequent generation of cells. There are two cell lines of inheritance, the **somatic line** and the **germ line**. **Somatic mutations** occur in the expanding cell population of the body (soma) but, by definition, somatic cells do not form the reproductive tissues. A somatic mutation might show itself as a harmless mole on the skin. Such a feature would have developed during one's growth and would therefore be absent from an identical twin. Such a localised distinctive feature would not be inherited. All of us experience somatic mutations and they accumulate as we grow older; they are one of the causes of ageing. Most are not harmful but they are degenerative, causing our body processes to slow down. Fortunately the human ovary and testis are set aside early in development and may not be so subject to accumulating mutational change. **Germ mutations** occur in these germinal tissues, the ovary and testis, that produce the germ cells, or gametes. Changes here may be passed to the next generation. It is this type of mutation that concerns us principally in this chapter, for being inherited they may be the cause of **hereditary disease**.

At its smallest, a mutation is a single base change in one of the genes along a DNA strand. A typical chromosome may have 100 000 kilobases correctly aligned in coded sequence, yet a change in the position of just one out of the hundred million bases may alter an allele to a form which, when transcribed as mRNA and then translated into protein, may be defective in its action. Such a mutant allele may initiate an hereditary disease in the new bearer of the defective DNA. Mutation can of course be on a much larger scale. There may be losses, gains or rearrangements of long lengths of the base code in one chance event. At meiosis, errors in chiasma formation (crossing over) may duplicate genes or cut them out, or even turn them back to front. Even whole chromosomes may be lost or gained, with considerable implications for the life of the individual concerned.

7.2   Mutation as an evolutionary clock

Not all mutations are bad news. A mutational change that occurs without demonstrable effect is called a **silent mutation**. Such changes only appear to the eye of the geneticist, who 'sequences' the DNA, recording the subtle differences that exist between the DNA of different people, and indeed between humans and our nearest relatives, the great apes. All animals have the same kinds of proteins functioning in their cells, but the greater the evolutionary distance between them, the more the sequence of the amino acids in the protein will differ, reflecting numerous small and often silent mutational events that have occurred over millions of years. Table 7.1 shows the number of base substitution differences between the DNA of humans and five different mammals in the code for the α and β chains of the haemoglobin protein molecule.

Table 7.1 The DNA base code differences in two haemoglobin chains between man and five other mammals.

	alpha (α) chain	beta (β) chain
chimpanzee	0	0
gorilla	1	1
African vervet monkey	5	10
mouse	19	31
rabbit	28	16

It is believed by some scientists that so steady is the way that these silent mutational events are picked up in the course of evolution that the constancy of rate maintains a molecular 'clock' that keeps time (Wilson and Sarich 1967). Using this method for the blood serum protein albumin, the clock has been 'set' to the divergence of the Old World primates from the New World primates at 35 million years ago. On this basis the gibbon line left the human line at about ten million years ago, the orang-utan at eight million years ago and the chimpanzee and gorilla at four million. Such figures seemed out of place with the fossil evidence when first announced, but are now thought to be in keeping with the dated geological evidence. Besides haemoglobin, six independent proteins have now been studied and each tells the same story. Much more recently direct comparisons of different species' DNA has been possible. If the 3×10^9 nucleotide sequence of human DNA is 'hybridised' with that of other primates to check the similarity of base sequencing, it is found to be 98.8% identical with that of the chimpanzee, 98.6% the same as the gorilla, 97.6% the same as the orang-utan and 93% the same as African monkeys. (See section 8.7 on DNA hybridisation.) Biochemically we are very close to our animal relatives but the distance between us has been increased by accumulated mutations.

Although many small mutational changes are silent it does not take much to cause harm. Mutational events are almost all for the worse. For humans some are mildly handicapping, some debilitating and some lethal. If harmful they are selected against and 'weeded out' in the struggle for existence. Their possessors survive less well. In the Neo-Darwinian view it is only the extreme minority of mutations that benefit their carriers. These are selected for; their possessors survive to reproduce multiple copies of their more successful DNA through the offspring they leave. Such changes of course may eventually alter a species; we should not forget that mutations provide the raw material for evolution.

7.3   The frequency of mutation

Individual genes mutate with a fairly fixed and regular frequency in the absence of outside influences. DNA replication errors are in the order of one mispairing of bases in a thousand million. As each gene has about 3000 bases, gene mutation rates are likely to be no more than about 1 in 350 000 for every copy made. Mutation rates for certain genetic conditions may be calculated from the rate that they appear and as such are usually expressed as the number of mutations per gene per gamete. Human mutation rates cover a big range from approximately 1×10^{-4} at the most frequent to 4×10^{-6} at the least. This is a range in chances from 1 in 10 000 to 1 in 250 000 gametes having a newly mutant allele at one particular locus. This rate of failure is of course made less serious by the fact that alleles are paired after fertilisation. As the great majority of new gene mutations are recessive, the chances of two alleles being newly mutant at one gene locus is infinitessimally small. Where new gene mutations are likely to make themselves immediately felt is where they are X-linked in a male or dominant in their expression.

Achondroplasia is due to such a dominant allele which is perhaps the best known cause of dwarfing in height (see Appendix to Chapter 7). All achondroplasic people have only a single dwarfing allele, for it is lethal in the homozygous condition. Heterozygote dwarfs may be born from one or two affected parents, but because the dwarfing allele is dominant in its expression, all such children born to normal height parents represent a new mutation of the gene. This makes it possible to calculate the mutation rate as being 14 per million gametes. 80% of people of restricted growth are born to normal height parents.

Duchenne muscular dystrophy (DMD) is a fatal genetic affliction of boys (see Appendix to Chapter 7), the exact cause of which is uncertain. The DMD gene mutates at a rate of about 8×10^{-5} per gamete. This is a frighteningly high rate and one in 4000 boys born in the U.K. suffers from this fatal condition, about one third of whom are the result of new mutations, whilst two thirds are born because their mothers are carriers.

Table 7.2 The frequency of new mutations

Condition	Type of inheritance (Autosome/X; Dominant/Recessive)	Rate of mutation (genes/ million gametes)	Chance of new mutation
Neurofibromatosis	AD	100	1 in 10 000
Duchenne muscular dystrophy	XR	80	1 in 12 500
Haemophilia A	XR	45	1 in 22 000
Achondroplasia	AD	14	1 in 70 000
Huntington's chorea	AD	5	1 in 200 000
Retinoblastoma	AD	5	1 in 200 000

The term 'mutant' may be correctly used for an individual or gene that first shows such a genetic change. As we are all probably possessors of new mutations, even though unexpressed, it is unhelpful to refer to somebody with an inherited disease as a 'mutant'. Again as the term 'disease' in popular language has implications of contagious character, many people today prefer to call hereditary conditions of this kind 'inborn errors of metabolism'.

7.4   Gene mutation

There are two types of gene mutation, **base pair substitution** and **frameshift mutation**. In the first of these if the A:T (adenine with thymine) and G:C (guanine with cytosine) base pairing is not faithful, that is one nucleotide pairs with the wrong partner, then at the next replication, where pairing is faithful again, a mutation will have occurred. This may happen if there is just a slight shift in electrons or small chemical change to the base. Sickle cell anaemia is a genetic disorder with this original cause. In this case a DNA mutation from thymine to adenine results in a mRNA change from GAA to GUA. This leads to an amino acid change from glutamic acid to valine in the β-chain of the haemoglobin molecule.

In frameshift mutation, the second type of gene mutation, a loss or gain of a number of base pairs (which is not a triplet code multiple of three) will cause the 'reading frame' of the ribosome to be shifted and all subsequent triplets will be mistranslated. Such an example of 'mis-sense' is the type of haemoglobin called Wayne. This has one chain five amino acids longer as a result of the deletion of a single base. If no functional protein results from the frameshift mutation it will be 'non-sense'. As three out of the 64 possible codon triplets (UAG, UAA or UGA) read as 'stop' signals, the chances of the protein being cut short are greatly increased by any frameshift mutation. The 'non-sense' code will most probably result in an enzyme that will no longer function. This is most dramatically shown by **albinism**. There are several stages in the synthesis of melanin, the black pigment of the skin. Normally the melanocyte skin cells take up the amino acid tyrosine and convert it to melanin in enzyme catalysed steps.

$$\text{Tyrosine} \xrightarrow{e^1} \text{DOPA} \xrightarrow{e^2} \text{DOPA quinone} \xrightarrow{e^3} \text{Melanin}$$

A failure of a cell to produce just one of these enzymes may cause albinism (see 6.4).

7.5   Chromosome mutations

Chromosome mutations are of two types, **structural aberration** and **aberration of chromosome number**. The latter is termed **aneuploidy** and is dealt with in section 7.6. Structural aberrations are associated with errors at chiasma formation, whilst errors of chromosome number result from errors in cyto-kinesis at cell division.

Chromosomes that come together in homologous pairs at meiosis match up in parallel and make chiasmata between their own and partner chromatids. Four basic types of structural aberration that may arise at this stage are recognised. Their possible origins are illustrated in Fig. 7.1. If the aligned pair of

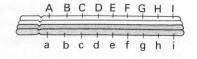

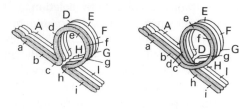

INVERSION

Inversion is caused by chiasmata forming across a loop within one chromosome

Cross overs between the chromatids of partner chromosomes, in this situation, may lead additionally to deletions and duplications of genes.

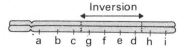

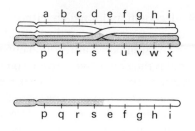

TRANSLOCATION

Translocation is caused by non-homologous chromosomes exchanging chromatids when mispaired at meosis prophase I.

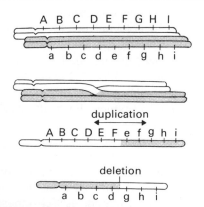

DUPLICATION AND DELETION

Misaligned homologous chromosomes which form chiasmata between the wrong loci result in both duplication and deletion of gene loci.

Fig. 7.1   The four types of structural aberration in chromosomes

chromosomes are looped then a chiasma *within* one chromosome of the pair may result in the loop becoming an **inversion**. Non-homologous chromosomes pairing in error may easily produce **translocation** of a part of one chromatid onto part of another non-homologous chromatid as a result of a cross-over. The most common types of chromosome mutation involve both **deletion** and **duplication**. These are both products of misaligned pairings of homologous chromosomes. Thankfully many such mutations are so damaging that they cause the death of the bearer before birth (natural abortion) or, more sadly, in early post-natal life. Naturally aborted foetuses are known to have a very high proportion of chromosomal abnormalities. Only one child in 1600 is born with detectable chromosome mutations of this sort. These individuals are often seriously physically handicapped or mentally retarded and many do not live out a normal life span.

7.6   Aneuploidy

The term 'euploid' means having a good or true set of chromosomes. The haploid number of chromosomes in a human gamete (23) or the diploid number in a normal body cell (46) would be described as **euploid**. Very rarely in animals, but much more commonly in plants, mutant multiple sets of chromosomes occur. This is termed polyploidy. Although it is known in humans it is very rare and invariably lethal.

Aneuploidy is an upset of the euploid condition in which either extra chromosome additions, or individual chromosome losses from the individual's normal complement of 46, take place. For any numbered autosomal chromosome pair a person is normally **bisomic**, that is they have two chromosomes in an homologous pair. If the cell that gave rise to the original egg cell or sperm cell was abnormal, the individual may have one extra chromosome or one less. Such individuals are called **trisomic** $(2n+1)$ or **monosomic** $(2n-1)$ respectively. A careful count of the chromosomes in a special microscope slide preparation will show up trisomy (47 chromosomes per cell) or monosomy (45 chromosomes per cell), for a matching up of the observed chromosomes into their pairs will show just which one is extra or is missing. (See 8.6 where analysis of chromosomes is described.)

Aneuploidy occurs by **non-disjunction** of homologous chromosomes at the first or second divisions of meiosis. When homologues or chromatid pairs separate normally (disjunction) one member of each pair goes to either end of the cell (see Fig. 3.1). However, if the spindle fibrils fail to pull the centromeres apart, then both may go to the same pole. (See Fig. 7.2) Gametes that experience non-disjunction in their formation will be haploid aneuploids $(n+1$ or $n-1)$. After fusion with another normal haploid gamete the resultant zygote will become a diploid aneuploid $(2n+1$ or $2n-1)$.

With the exception of the X chromosome, aneuploid monosomics missing one autosomal chromosome $(2n-1)$ do not survive for long even in foetal life. Aneuploid trisomics having one extra chromosome $(2n+1)$ survive much better, most until after birth. By far the most common and with the longest survival chances is **Trisomy 21**, in which there are three copies of chromosome 21. This condition, known today as **Down's Syndrome**, was once called 'mongolism', a name that should not be used. In the recent past it has occurred as commonly as 1 in 700 live births. (See Appendix to this chapter.)

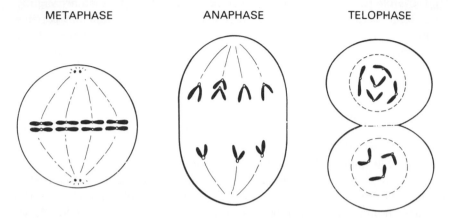

METAPHASE   ANAPHASE   TELOPHASE

Fig. 7.2 Non-disjunction during any cell division may result in daughter cells having one more or one less chromosome. In this example there are four chromosomes only. Where non-disjunction occurs in the meiotic divisions preceding gamete formation the cells of the offspring arising from fertilisation will have either one chromosome extra or one less.

Sex chromosome aneuploidy occurs at a rate of about 1 in 300 live births, a surprisingly high rate. Most of these individuals will be quite unaware of their mutation and will not manifest any marked abnormality. However, those with **Turner's Syndrome** and **Klinefelter's Syndrome** will be affected both physically and socially. For this reason these humanly important conditions are described in the appendix to this chapter. Sex chromosome aneuploidy has its origins in the non-disjunction of the sex chromosomes in egg and sperm formation. Abnormal ova may carry two X chromosomes or none at all. Abnormal spermatozoa may carry XY or no sex chromosomes, as a result of non-disjunction at the first division of meiosis, or XX or YY as a result of non-disjunction at the second division of meiosis. These aneuploid gametes coming together with normal gametes may produce four commonly found combinations. (See Table 7.3.)

Table 7.3 Sex chromosome aneuploidy

Genotype	Sex	Condition	Origin	Incidence at birth
XO	♀	Turner's syndrome	X sperm + O egg O sperm + X egg	1 in 2500
XXY	♂	Klinefelter's syndrome	Y sperm + XX egg XY sperm + X egg	1 in 1000
XXX	♀	normal	XX sperm + X egg X sperm + XX egg	1 in 1000
XYY	♂	almost normal	YY sperm + X egg	1 in 1000

Triple X women seem to be entirely normal. Their condition only comes to light as a result of chromosome analysis. Hardly any of their sons are Klinefelter males, while perhaps one half would be expected to be. XYY men are slightly taller than average, are slightly less intelligent than average, and may possibly be more aggressive in temperament. The incidence of this condition is 1 per 1000 male births, but there are reputedly 20 per 1000 such males in penal institutions for the mentally subnormal. Such prisons inevitably contain individuals who are more violent and less able to adjust socially. Should this affect our general attitude to XYY males? Certainly it would be quite wrong to regard them as 'criminals'.

7.7   Genetic diseases

Diseases that have a cause related to an individual's genetic inheritance are perhaps more obviously seen today now that the pathogenic diseases caused by micro-organisms are under better control. Although some conditions such as colour-blindness, albinism or dwarfism are a mild disability or social handicap, many disorders have a more disease-like effect and may be distressing to all concerned. Too often they are seen in children and those children may die very young, even after the most loving care and attention has been lavished on them. The parents of those with genetic diseases are told the cause by their doctors. The parents very often do not really understand, they feel guilty because of the suffering they seem to have caused their children and they may well wonder about environmental hazards to which we are all exposed. The doctors feel relatively helpless because the causes lie deeper than their medicines can reach. All of this presents a challenge to our society which today we should be better able to meet. The next chapter takes up this challenge of genetic disease. The appendix to this chapter gives an outline of the most important disorders dealt with by this book.

Appendix to Chapter 7

Autosomal gene mutations

Achondroplasia

Inheritance: Dominant autosomal
Incidence at birth: 1 in 26 000
Mutation rate: 14 per million gametes

Achondroplasia is just one form of dwarfism. Dwarf people, who prefer to be known as people of restricted growth (mean adult height 127 cm) are able to lead a fully normal life. All are genetically heterozygous, for the homozygous condition is lethal. The trunk is of normal length but the limbs are very much shortened. The forehead has a deeply depressed nasal bridge. Fertility between dwarfed parents is reduced. All achondroplasics are of normal intelligence. Although many suffer physically from back-ache they often suffer far more socially, due to the bigotry of the rest of society.

Albinism

Inheritance: Recessive autosomal at two gene loci
Incidence at birth: 1 in 40 000 (Europe)

Albinos most commonly lack the enzyme tyrosinase and are thus unable to synthesise melanin. The skin is pink and fails to tan, the iris is pale blue or pink and the eyes easily reflect red from the unpigmented choroid, in a way that only a flash photograph will show with a normal eye. The hair is pure white. Albinos suffer from sunburn but have no other particular medical problems beyond a difficulty seeing in bright light. Almost all albinos have heterozygous normal parents. Albinos can easily suffer from ill-informed social attitudes. Albinism is perhaps the earliest recorded human genetic condition. According to the Book of Enoch, Noah was an albino

'. . . a child whose flesh was as white as snow and red as a rose, the hair of which was white and long and whose eyes were beautiful. When he opened them he illuminated all the house like the sun'.

Noah's parents, it is recorded, were first cousins.

Cystic fibrosis (CF)

Inheritance: Autosomal recessive on chromosome 7
Incidence at birth: 1 in 2500 (Northern Europe)

Cystic fibrosis is very common in Europeans and has the highest incidence of any single gene hereditary disease in Britain (one person in 25 is a carrier). The homozygous recessive CF sufferers are generally children with most commonly two normal but heterozygous parents. They produce an abnormally thick mucus in all mucus secreting tissues, such as the respiratory passages and the gut. They also have a characteristically high sodium excretion level in their sweat. The CF gene is thus pleiotropic (see 6.4) for it causes several phenotypic effects. There is chronic lung congestion that may be relieved by physiotherapy. Antibiotic treatment is needed to guard against infections. There is pancreatic obstruction and trypsin deficiency, and liver damage and diabetes commonly ensue. Some patients survive well but many children die young; there is no cure. Median survival is 19 years. The location of the CF gene is precisely known. There is, or in the past must have been, an advantage to heterozygotes, to explain the high gene frequency (see also sickle cell anaemia). The gene has been isolated by genetic engineering (1987) and soon it may become possible to detect all carriers. Analysis of amniotic fluid for foetal enzyme levels identifies affected foetuses at 18 weeks with 95% confidence.

Huntington's chorea (HC)

Inheritance: Autosomal dominant on chromosome 4
Incidence: 1 in 18 000 in the U.K.
Mutation rate: 1 in 200 000 gametes

In 1872 George Huntington, a U.S. doctor, first described this truly frightening nervous degeneration disease. The onset of the illness is usually in early middle age and death occurs within ten years. There is no cure. Most commonly a sufferer will have had one parent die of the condition. The mutant allele is dominant, thus potential sufferers, who will be heterozygous, have a fifty per cent chance of passing the allele on to each of their children before they themselves realise that they are afflicted. Knowledge of the condition in a family creates psychological strain. Fortunately the disease is rare. Prenatal diagnosis is under intensive research investigation. There is every chance that with the new gene technology and with genetic counselling this condition may have a much reduced incidence in the future.

Phenylketonuria (PKU)

Inheritance: Autosomal recessive on chromosome 12
Incidence: 1 in 10 000 in the U.K.

PKU is a common hereditary disease but since widespread screening of newborn babies in the U.K. (begun in 1961) its worst effects have been reduced. PKU sufferers lack the enzyme phenylalanine hydroxylase (PAH) in the liver. Phenylalanine is an amino acid. If an excess of this amino acid is in the diet the molecules are converted to toxins that severely limit mental development. Mental subnormality is characteristic unless special diets are followed. Because phenylalanine is on a metabolic pathway to melanin synthesis, PKU sufferers are often very fair-haired. If reared on a phenylalanine re-stricted diet from birth there is virtually no mental handicap. The PKU gene has been cloned. Somatic gene therapy by genetically engineered cell implant is a real possibility. Carrier detection is possible.

Sickle cell anaemia

Inheritance: Autosomal codominant
Incidence: 1 in 400 000 in U.K., 1 in 400 U.S.A. black population
1 in 40 in worst affected parts of East Africa

Sickle cell anaemia (sickle cell disorder) is a disease associated with human populations living in malarial areas. The highest incidence is in regions of East Africa (4% of births). In the U.K. it is most frequent amongst blacks who can trace their ancestry back to the West Indies and to the highly malarial areas of West Africa. One hundred thousand sufferers are born annually in Africa; about 140 are born in the U.K. The characteristic sickling of red blood cells is detectable in heterozygote carriers. These otherwise normal individuals who show what is called 'sickle trait' have a high resistance to malaria and hence have a higher survival chance and reproductive fitness than non-carriers in malarial areas. Those sufferers homozygous for the allele have a severe haemolytic anaemia with onset in infancy. Infant mortality is very high. Distorted red cells block capillaries causing a cessation of blood supply. The spleen is destroyed and the sufferer is prone to infections. Regular blood transfusions can lead to a near normal life. Carrier detection is simple, by the electrophoretic separation of normal haemoglobin (Hb^A) from the abnormal (Hb^S). Couples of African ancestry can easily be screened at a genetic counselling clinic in the U.K. or U.S.A. Hb^S is not produced by the foetus, and hence foetal blood samples cannot be used for an early diagnosis, though recent progress has been made with the use of gene probes (see 8.7). Some advances in the treatment of sickle cell disorders have recently been made.

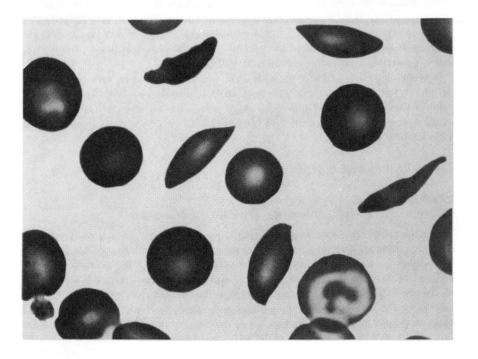

Fig. 7.3   A few normal, disc-shaped red blood cells surrounded by distorted sickle cells in the blood of a person with sickle cell anaemia

Thalassaemia

Inheritance: Autosomal recessive on chromosome 11.
Incidence: Rare amongst most of U.K. population, but as high as 1 in 130 amongst Cypriots and 1 in 800 amongst Greeks and Italians.

Thalassaemias are due to a reduced or imperfect synthesis of the haemoglobin α and β polypeptide chains. Both conditions are, like sickle cell, associated with populations that have adapted to malarial areas (see 10.5). Beta thalassaemia is the most likely form to be encountered in the U.K. and is due to a variety of gene mutations, such as premature STOP signals, frameshift and deletion. Heterozygotes have almost no symptoms of anaemia and are essentially normal, but may be identified by routine laboratory tests. Homozygote sufferers have normal foetal haemoglobin but in infancy rapidly deteriorate in health as they cannot make haemoglobin properly. Some amelioration is possible with blood transfusion but the condition is effectively lethal.

Sex-linked gene mutations

Haemophilia A

Inheritance: X-linked recessive
Incidence: 1 in 5000 male births

The blood clotting protein Factor VIII, also called anti-haemophiliac globulin (AHG), is either defective or reduced in amount. Haemophiliacs therefore suffer internal or external bleeding easily, especially into their joints. The condition is totally contained by regular injections of human AHG. This was, in the U.K., obtained very largely in blood products from the U.S.A. before the present heat-treatment was made compulsory. Sadly, many U.K. haemophiliacs are now HIV positive.

Haemophiliacs who survive to adulthood and father children will have carrier daughters, but not afflicted sons. Sisters of haemophiliacs have a 50% chance of being carriers. Carrier detection is improving and genetically engineered AHG may soon be available in a clinically tried form. 90% of haemophiliacs are born from carrier mothers, 10% are new mutations.

Duchenne muscular dystrophy DMD

Inheritance: X-linked recessive
Incidence: 1 in 4000 male births, one third are new mutants

Duchenne (1868) first described this exclusively male disease. Muscle weakness and uncoordination are detectable in infancy as the small boy shows signs of being unable to climb stairs or walk evenly. The legs are spread widely and the posture hunched forward. Muscle is replaced by fibrous tissue and the child is wheelchair-bound by the age of twelve. Muscle wasting continues remorselessly until even breathing is impossible. These young men, whose families have given them much physical and emotional support, usually die from pneumonia before they are twenty. The DMD gene has been precisely located on the short arm of the X chromosome; it codes for a single very large protein of uncertain function. Much research is being done on the detection of carriers and on pre-natal diagnosis.

Aneuploid chromosome mutations

Down's Syndrome

Inheritance: Trisomy of chromosome 21, due to non-disjunction
Incidence: 1 in 700 births, at mean age of childbearing

J. Langdon-Down (1866) first described this condition clinically. It was for long known in Europe as 'mongolism', because of the slanting eye and flattened face features, and was attributed in folklore to genes left behind in Europe by the invading mongol armies. The term 'mongolism' is insulting to anyone from Eastern Asia (where incidentally children with Down's Syndrome are considered to show European features!) The trisomic cause was identified in 1959 by J. Lejeune. A child with Down's Syndrome has a short broad skull, short neck and distinctive face with prominant eyelids that slant down to a short nose. They have short stature and the tongue often protrudes. The iris has spots around the pupil. The hands are short and broad with a simian crease across the palm. Individuals with Down's Syndrome that survive through infancy, when heart and lung defects may well occur, have a near normal lifespan. Much controversy surrounds the depressed intelligence of children with Down's Syndrome. IQ scores are generally below 50 (see Chapter 9) but if much care is invested in these children they are commonly able to achieve near normal social and intellectual growth and, importantly, personal self-sufficiency. The children certainly repay, in a most rewarding way to those around them, the affection that they have received. (It is sad that in our society we esteem people more by their heads than their hearts!)

There is a maternal age effect in Down's Syndrome (see Fig. 7.4) nearly 80% of the non-disjunctions being of maternal origin. Primary oocytes that have been in the mother's ovary for more than 30 years seem more prone to non-disjunction at meiosis. Paternal non-disjunction explains 20% of cases but is not so age-related. Foetal chromosome analysis in pregnancy is often advised if a mother is over the age of 35. Because mothers carrying a foetus with trisomy 21 often elect for an abortion, the proportion of children with Down's Syndrome born today to younger mothers is rising, although the absolute number born is falling each year. Half the children of mothers with Down's Syndrome will inherit the condition.

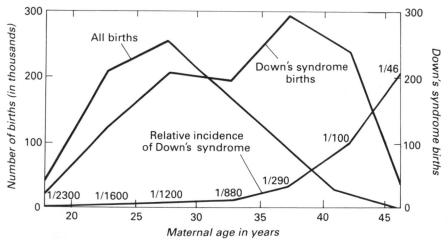

Fig. 7.4   The maternal age effect in the incidence of Down's syndrome. Down's syndrome births divided by all births gives the relative incidence of occurrence. (USA data prior to 1980, taken from Jenkins 1983)

Turner's Syndrome

Inheritance: X Monosomy (XO) due to non-disjunction
Incidence: 1 in 2500 females at birth

This condition is very common at conception but is near fatal for the foetus, 99% being aborted. It presents no life-threatening problem after birth. The XO person is female (lacking the HY-antigen), is short in stature (mean 130 cm), with a wide neck, broad chest and relatively little feminisation. The ovaries do not develop, there is no pubic hair, no menstruation and complete infertility. Oestrogenic hormone therapy improves normal sexual development but does not increase stature. There is little mental retardation. Contrary to the origin of Down's Syndrome, Turner's Syndrome usually arises from meiotic non-disjunction during spermatogenesis, but there is no comparable paternal age effect apparent.

Klinefelter's Syndrome

Inheritance: XXY Trisomy, due to non-disjunction
Incidence: 1 in 800 births

First described by H. F. Klinefelter (1942) this condition was identified as XXY trisomy in 1959. Individuals are male with some more feminine features that few people would notice. The limbs are elongated from early childhood, the testes are small and male secondary sexual charcaters weakly expressed. There may be some breast development. The Klinefelter male has a low sex drive and is most often infertile. Male hormone therapy will help increase the male characteristics and perhaps self-esteem. Klinefelter individuals are below average intelligence; 20% are mildly mentally retarded.

8   The challenge of genetic disease

8.1   Introduction

Diseases are of many kinds; we may recognise ones caused by pathogenic micro-organisms, by parasites, by malnutrition or by bad physical or social environments. For any such disease, recognition of the cause has been the important step in seeking a cure. In the past hereditary diseases have been both seemingly incurable and inexplicable. Today they are no longer difficult to comprehend and are conditions which science and medical technology can grapple with much more effectively. As medicine has eliminated so many diseases of other kinds, the proportion of illness attributed to genetic causes must clearly increase. As medical and nursing care also improve, a higher and higher proportion of those in our society who have poor health are seen to have a genetic basis for their illness.

Today we view health as being more than an absence of disease. Part of healthy living is being more aware of the way in which our bodies operate. Doctors increasingly welcome the situation in which they are seen to be advisers and attentive counsellors with medical knowledge, rather than as distant and aloof prescribers of pills. As their patients we will be the better if we steer ourselves towards health under their guidance. To meet the challenge of genetic disease we need much new knowledge, and more compassion and real care for the sufferers. We also need a considered attitude to the usefulness and value of new medicine and biotechnology. Knowing what it is *right* to do requires that we have answers to ethical and moral questions; some of these will be examined in Chapter 11. This chapter is concerned with a better understanding of the causes of genetic damage; how we can know where the disorder is in the genome; how the disorders are diagnosed; and how people may come to terms with genetic illness in their own lives and that of their families.

8.2   Environmentally induced mutation

There is a naturally occurring rate of spontaneous mutation that may be accelerated by specific environmental agents. Such **induced mutations** are caused by ionising radiations and particular chemical substances that are together called **mutagens**. To a large extent there are no differences in kind between natural and induced mutations, other than that the mutation rate is accelerated and the increase of induced mutations may be associated with a particular environmental agent. Disorders of cellular growth, to form tumours or cancers, often arise because of genetic change. Chemicals and radiations that are **mutagenic** are often also **carcinogenic** (giving rise to cancer). Because exposure to mutagens is harmful to genetic material, an avoidance of mutagens is sound **primary genetic health care**.

Mutagenic chemicals

Some of the chemical substances we meet daily, in what we eat or inhale or touch, may be mutagenic or carcinogenic. Many of these are de-toxified by the body, quickly excreted, or repairs are made by our cells to the damage that they cause. Mutagens that are particularly powerful commonly attack the DNA coding system directly. **Mustard gas**, first used in the 1914–18 War and most recently in the Iran–Iraq war, is one such mutagen. Besides causing burns to the skin, eyes and lungs it brings about base pair substitutions in the DNA of cells it reaches. **Bromo-uracil** is a similar mutagen which displaces the base thymine from its position and then subsequently pairs with guanine at a later replication. In this case an A–T pairing becomes a G–C pairing and the code meaning may be broken. (See Fig. 8.1.)

Many common compounds are mutagenic in large concentrations or under special circumstances; these include vinyl chloride (in plastics), trichloro-ethylene (a commonly used solvent), saccharin (a sweetener), sodium nitrite (a meat preservative) and caffeine (a stimulant in tea and coffee). Some pesticides and many herbicides kill pests and weeds by damaging their nucleic acids. For this reason commercially grown vegetables should be washed before being consumed. Tobacco smoke contains many mutagens of the polycyclic aromatic hydrocarbon type. Some hair dyes have been found to be frameshift mutagens. These are absorbed through the scalp when applied to the head and then

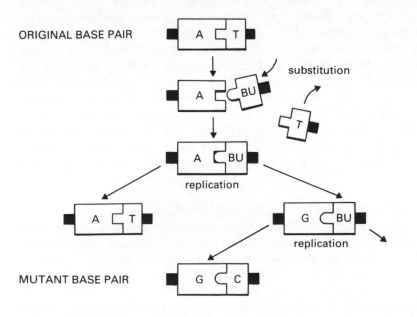

Fig. 8.1   Gene mutation by base pair substitution (see text)

largely excreted by the kidneys. About 30 000 new chemicals are put onto the market in industrialised countries each year. Although many are beneficial to us and they are generally stringently tested, new mutagenic hazards are always likely to be there and must be guarded against.

Damaging radiations

High energy radiation which may be thought of in the form of electromagnetic waves or tiny particles is known as **ionising radiation** (wavelength 10^{-7} to 10^{-12} metres). When DNA is irradiated it may undergo chemical change or literally be cut by the particle impact. Ultraviolet rays, in sunlight, move electrons out of their normal orbits, chemically changing the DNA and often killing cells. The incidence of skin cancer amongst habitual sunbathing white people, who are not protected by black melanin that absorbs UV, is well known. Although UV-rays do not go far into the skin, X-rays and gamma rays may pass right through the body. High dosages of these radiations to the testis and ovaries is certainly potentially damaging to offspring. Radioactive caesium, iodine and strontium are amongst the 'fallout' components from nuclear tests in the atmosphere or from reactor accidents (such as that at Chernobyl in April to May 1986). As chemical substances, such radio-isotopes are absorbed into living organisms and pass into the food chain. These are beta particle emitters and may do tremendous damage over a short distance in the tissues in which they come to reside. There has been much research into the damaging effects of radiation. The following are key conclusions:

1. A wide range of mutations are formed, particularly in actively dividing tissues (skin, bone marrow, ovary, testis, gut lining) or in well-oxygenated tissues (the brain).
2. The rate of gene mutation is proportional to the dose. There is no 'safe level' of radiation, though there may be a legally drawn 'acceptable level' (which may be well below that in some natural environments!).
3. At low intensity and with long exposure the radiation will cause damage to DNA that is more effectively repaired than if the same dose is given at higher intensity with a short exposure.
4. Ovary and testis seem to screen out major chromosomal anomalies. The same function is served by natural abortions.
5. Gene mutations that are radiation induced may be inherited.

There are numerous non-natural radiation sources to which we are exposed. These represent a fraction of 'background radiation' which in some natural environments is in fact dangerously high. High level short duration radiation is devastating in its biological effects. 46 000 people who were children in Hiroshima and Nagasaki in 1945 have been monitored since. There has been a definite shift in the sex ratio of their children to substantially more females than males being born. This presumably reflects an increase in X-linked recessive lethals. The closer these individuals were to the explosions, the higher the chromosomal damage. First cousin marriages between the grandchildren of those exposed people may show up many of the gene effects that were induced in 1945. In the ensuing months and years cancers were very prevalent amongst those who had been irradiated.

Many scientists regard nuclear weapons preparations and nuclear power generation as environmentally threatening 'health hazards'. Each of us should assess these radiation health risks on the basis of sound biological knowledge and the other factors against which any risk that we take is weighed. Radiation is by no means all 'harmful'. Radioactive isotopes and beamed damaging radiations play an important part in cancer therapy, in disease diagnosis and in X-ray radiography.

The repair of DNA

Knowledge of the way that cells mend their broken DNA may be of great significance for future treatment of genetic diseases. Recent evidence points to DNA, RNA and protein synthesis all being under continued cellular surveillance for abnormalities. In DNA, mismatched pairings of bases across or even along one chain of the helix are removed by healthy cells. The damaged portion is cut out by a **binding protein** and the gap then made good by **DNA polymerase** and **ligase**. (See Fig. 8.2.) Understanding such repairs may one day make gene insertion by a type of microsurgery a possibility.

In the genetic disease *Xeroderma pigmentosum*, malignant skin cancers called **melanomas** develop much more frequently in response to UV light. For such people the disease may be due to a lack of the necessary repair systems that the body normally possesses for dealing with such potentially cancerous situations. There are a number of **chromosome instability diseases** where major chromosomal repair systems are defective. For these affected individuals, who may be more prone to cancer, there is probably a lack of the enzymes that normally keep our genetic material in disciplined order.

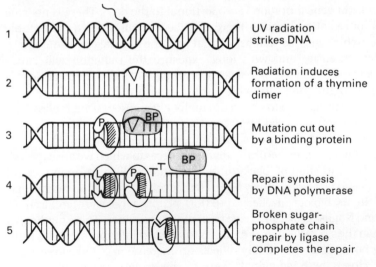

1   UV radiation strikes DNA

2   Radiation induces formation of a thymine dimer

3   Mutation cut out by a binding protein

4   Repair synthesis by DNA polymerase

5   Broken sugar-phosphate chain repair by ligase completes the repair

Fig. 8.2   Stages in the repair of DNA

8.3 The frequency of defective alleles in a population

There are very large numbers of defective gene conditions (2800 recorded in 1986) but why are so relatively few expressed and what proportion of people are carrying them? These are questions that can be answered.

It should be clear that single gene mutants are often unexpressed in their carriers because the vast majority are recessive alleles masked by a dominant functioning partner allele. Diploidy, the possession of chromosomes in pairs, is a sort of insurance policy. Where there is a single chromosome, as is the case with the X chromosome in males, there is markedly more expression of the X-linked defective alleles in such XY individuals (see references to haemophilia, Duchenne muscular dystrophy and colour-blindness).

To understand the **frequency** with which a defective allele is likely to turn up, we need to think about the population, not of people, but of different chromosomes in the population as a whole. If we were to collect all the chromosomes of one kind, like all the X chromosomes, what would be the frequency of one type of allele as opposed to another at *each* gene locus? If we suppose that for one gene locus on the X chromosome, one allele in ten is defective, we would say that its **allele frequency** would be $\frac{1}{10}$ or 0.1. This would mean that a tenth of all men would express the condition, because men only have one X chromosome and one tenth of them, on average, would have an X chromosome with this defective gene present. As women have two X chromosomes their chance of being homozygous for the condition would be $\frac{1}{10} \times \frac{1}{10}$ or $\frac{1}{100}$ ($0.1 \times 0.1 = 0.01$), a far smaller proportion.

Most inheritance involves chromosomes in homologous pairs; thus where an autosomal homozygous recessive condition does occur, its frequency in the population will be much less than the frequency of the allele in the whole gene pool. Such genotype and allelic frequencies are of course related, for the **genotype frequency** is the square of the allelic frequency. This is shown by the square diagram Fig. 8.3 for two alleles of differing frequency that combine together in pairs.

In Fig. 8.3 there are two alleles A and a. The portions of the sides of the square over which they extend are proportional to their frequency in the population as

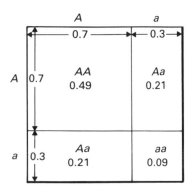

Fig. 8.3

a whole. The relative sizes of the boxes within the square represent the genotype frequencies that result from their combination within the population as a whole. The two allele frequencies sum together to 1.0. The individual genotype frequencies AA, Aa, aA and aa also add up to 1.0.

The genotype frequency of AA $= 0.7 \times 0.7$ $= 0.49$
The genotype frequency of aa $= 0.3 \times 0.3$ $= 0.09$
The genotype frequency of Aa and $aA = 0.7 \times 0.3 \times 2 = \underline{0.42}$

Total genotype frequency $\underline{1.00}$

Because the genotype frequency in the homozygote is the square of the frequency of the allele in question, the square root of the genotype frequency for the homozygous condition is the frequency of the allele. These mathematical ideas are represented algebraically by the **Hardy-Weinberg equations**. These were first put forward in 1908 by G. H. Hardy and W. Weinberg, quite independently, to express the frequency of alleles in a population as a function of the genotype frequency observed. Where p and q represent the frequency of the alleles, the gene frequency is expressed as
$$p + q = 1.$$
Because the frequency of the genotype occurring is the square of the gene frequency, the genotype frequency is
$$(p + q)^2.$$
The frequencies of the 3 genotypes also sum to 1.
Thus as
$$(p + q)^2 = (p + q) \times (p + q)$$
$$= p^2 + 2pq + q^2$$

Then the genotype frequency is expressed as
$$p^2 + 2pq + q^2 = 1.$$

This notion is easily expressed by the square diagram Fig. 8.4.

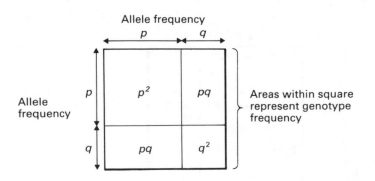

Fig. 8.4

If these two equations can be understood, in the simplest way, as expressions of the relationship between how many alleles there are around in the population, and how many people there are that show the features that those alleles express, so much the better.

The gene frequency equation $p + q = 1$

The genotype frequency equation $p^2 + 2pq + q^2 = 1$

8.4 Using the Hardy–Weinberg equations

There are many uses to which geneticists may put the Hardy–Weinberg equations, but the simplest and most important is to estimate what proportion of the population are carriers for a recessive genetic condition. One of the assumptions that we will initially have to make is that the frequency of the alleles does not change markedly by mutation, adding or subtracting alleles, or by natural selection acting to alter their possessors' chances of survival.

Let us take the example of albinism. One person in 40 000 is an albino with the genotype *aa*. Those with the genotypes *AA* and *Aa* represent the remainder of the population. What proportion of the whole population are (*Aa*) carriers of the recessive allele? These are the individuals who, if they married, would have a one in four chance of producing albino children. To work this out proceed as follows:

Let the frequency of allele $A = p$
Let the frequency of allele $a = q$
In the population as a whole the frequency of the genotypes *AA*, *Aa* and *aa* will be represented by p^2, $2pq$ and q^2,
where

p^2 represents the frequency of the *AA* genotype
$2pq$ represents the frequency of the *Aa* genotype
q^2 represents the frequency of the *aa* genotype

As q^2 represents the frequency of albinos (*aa*),

$$q^2 = \frac{1}{40\,000} = 0.000\,025$$

therefore

$$q = \sqrt{0.000025} = 0.005$$

As $p + q = 1$

$$p = 1 - q = 1 - 0.005 = 0.995$$

The frequency of *Aa* carriers is therefore
$2pq = 2 \times 0.995 \times 0.005$
$= 0.00995$
$= 0.995\%$ of people.

This indicates that approximately one person in a hundred is an albino carrier. They are therefore by no means rare. What is unlikely is for *two* such people to marry (1 marriage in 10 000) and to produce albino offspring (with a 1 in 4 chance). Similar calculations make it possible to estimate the theoretical carrier rates for a number of common genetic diseases. Because there are so many recessive gene defects, it is helpful to regard ourselves as each being carriers for some quite badly defective alleles. Whether they find expression in us or in our children must remain a matter of chance.

Geneticists use the Hardy–Weinberg equations, together with data on survival rates of those with handicaps and on their reproductive potential, to calculate the rates at which new mutations are appearing. This is the basis for the statements of actual mutation rate that are given earlier in this chapter and for the diseases described at the end of Chapter 7. In the table below some common carrier frequencies are given (U.K. data).

single gene disorder	carrier %	chance of being a carrier
Cystic fibrosis	3.9%	1 in 25
Phenylketonuria	1.9%	1 in 50
Alcaptonuria	0.62%	1 in 600
Sickle cell disease	0.3%	1 in 318
Haemophilia A	0.036%	1 in 2800 women
Duchenne MD	0.03%	1 in 3000 women

The knowledge that one person in twenty-five is a carrier of cystic fibrosis illustrates that the genes for these conditions are not 'rare'. We should be able to feel more sympathy for those who are unlucky enough to suffer from such conditions and be more committed directly or indirectly to their families.

8.5   The human cost of genetic disease

Most hereditary diseases come to light at birth or in early childhood. Today some are routinely checked for during pregnancy or at birth. A good doctor is always on the watch for them. Because other diseases have been so effectively reduced, genetic diseases have increased in relative prominence. They accounted for only 3% of the infant mortality rate in England and Wales, in 1900, but today they make up the stated cause of death for 40%.

It is important to realise that **spontaneous abortion**, or miscarriage, prevents many genetically handicapped children from being born at all. We know that 7.5% of conceptions show some chromosomal abnormality and 50% of stillborn children have deformities. It is perhaps as well that they should not survive given what we know of their genetic health. One third of childhood deaths have a genetic cause. Between 3 and 4 children per thousand are mentally retarded for genetic reasons and 10 to 20 per thousand have severe physical handicaps. Once these children are born into our society, few people question our duty to give them every care, but is it right to turn a blind eye to the causes if they can be anticipated early enough? The strains of caring for severely handicapped children do sometimes lead to great unhappiness, to divorce and to neglect of brothers and sisters. The lifetime financial costs of caring for a person with Down's Syndrome may be very considerable. Does this matter? Should we not accept the gift of human life and keep the child alive whatever its condition from the time of conception? How would our attitude be affected if the condition could be diagnosed long before birth? Is the **induced abortion** of a foetus known to be handicapped morally justified? (See the discussion of eugenics and euthanasia in Chapter 11.) Our society has at present conflicting moral views. Science can clarify some of the issues, but it is not equipped to make the rules for society.

Quality of life for any individual is the key to their happiness as people. Although many of the individuals that suffer from genetic diseases may be thought to be unhappy in themselves, this need not be. Not only are many able to lead very rewarding lives, but those who care for them may often find a vocation in the love that they are able to give. It is important that this be remembered.

8.6 The diagnosis of genetic disorders

Modern medicine has provided us with a vast array of techniques for discovering various disorders. These include whether a person is a carrier of certain genetic diseases, if they have a chromosome abnormality or if a foetus has a specific gene defect or a chromosomal disorder, (long before the end of pregnancy). Often it is the arrival of a handicapped child into a family that prompts a medical investigation. Before we consider the important task of the genetic counselling of people who want to know more about genetic disease, it is worth understanding the methods of diagnosis employed.

Foetal diagnosis

Parents who have already had a handicapped child will be anxious lest any subsequent children be similarly afflicted. Where the parents are at all prepared to consider the abortion of an affected foetus, they may be prepared to undergo some form of foetal diagnosis. Diagnostic methods are either 'invasive' or 'non-invasive' of the womb itself. The use of **ultrasound scanning** is a non-invasive method used for imaging. By converting high frequency sound waves into pictures on a TV screen it is possible to detect major deformities of the foetal brain, kidney, heart and limbs. There are believed to be no risks attached to the method, which is not so true when X-rays are employed. **Radiography** may be useful for detecting some bone deformities but increasingly ultrasound is preferred. Ultrasound is invaluable for guiding the use of a needle or probe in a more 'invasive' method.

There are three principal invasive techniques, fetoscopy, amniocentesis and chorionic villous sampling. **Fetoscopy**, which is best done at 18 to 20 weeks gestation, employs a fibre optic device inserted through the cervix, that enables foetal features to be examined in more detail, if it is considered to be worth the risk. Foetal blood may be carefully sampled at this stage and diagnosis of haemophilia and β-thalassaemia are possible.

Trans-abdominal **amniocentesis** may be carried out in the 16th to 18th week of gestation, and involves the withdrawal of amniotic fluid for chemical and chromosomal analyses or for DNA diagnosis. This has been the principal method for early detection of Down's Syndrome or other chromosomal defects and makes easy the sexing of the foetus. Of the 14 000 tests done per annum in

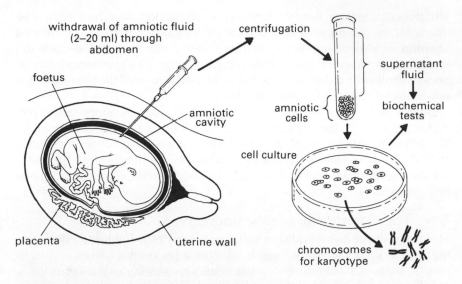

Fig. 8.5 Amniocentesis (the withdrawal of amniotic fluid) and the subsequent biochemical and cytological testing involved in foetal diagnosis.

the U.K. 2% have revealed abnormalities. Parents have chosen to have an abortion of the foetus in more than a half of these cases. Employing amniocentesis and fetoscopy slightly increases the risk of spontaneous abortion. This is 2.8% following amniocentesis, that is 0.3% above the natural rate (2.5%) at this stage of pregnancy. It is generally held that the added risk is worthwhile.

Chorionic villous sampling (CVS) or **chorion biopsy** is likely to replace amniocentesis as the major invasive diagnostic technique. It is done at 8 to 12 weeks of gestation when the embryo is only a few centimetres in length. This has the great benefit of allowing adequate time for the culture of cells and for any necessary decision about termination of the pregnancy before it is very advanced, as it will be after amniocentesis. The technique involves the insertion of a fine tube through the vagina into the uterus, under the image guidance of ultrasound. A minute sample of chorionic villi, the finger-like extensions of the early placenta, is taken for cell culture. This can then be used for chromosome analysis or DNA diagnosis. From the cells that are cultured more than 60 different biochemical disorders may be tested for. Whereas it takes only 1–2 weeks to do chromosomal analysis, enzyme studies may take 6–8 weeks (see notes on cystic fibrosis, Appendix to Chapter 7). The option of an abortion clearly becomes more traumatic if tests are to be time-consuming. Spontaneous abortion (miscarriage) at this early stage (8–12 weeks) is not uncommon, terminating some 7% of conceptions. CVS does not seem itself to add appreciably to the rate, and is the subject of much investigation at present.

Karyotyping; the analysis of chromosomes
Cells from amniocentesis or chorionic villous sampling are cultured **in vitro** for a few days. A sample of cells is treated with colchicine, a mutagen that arrests

cell division at mitotic metaphase by preventing spindle formation. The cells are then exposed to a dilute salt solution so that they swell osmotically and so space their chromosomes apart. They are then stained to give the chromosomes their characteristic banding pattern. (See Fig. 2.1) Fig. 8.6 displays in diagram form the full human chromosome complement, arranged in order of decreasing length, with the short arm uppermost. This is referred to as the species **karyotype**. The chromosomes are numbered 1 to 22, X and Y by an international convention (1971). The complex banding patterns that may be produced allow the precise identification of different gene loci.

Karyotyping is done by taking a photograph of the cell, cutting out the chromosomes from the print, pairing up the homologues and setting them out in numerical order. In this way numerical anomalies, such as Down's Syndrome, and translocations may be detected. Sometimes, for example, chromosome 21 may adhere to chromosome 16. This can result in three sets of chromosome 21 in a cell (trisomy) producing all the symptoms of Down's Syndrome in the bearer. Careful studies of family pedigrees often show an association between a genetic disease and another phenotypic condition. For example haemophilia A is very closely linked with colour-blindness, for the two gene loci are next to each other on the long arm of the X chromosome. Some single gene defects are known to be on a particular chromosome, because where they occur it is possible to detect on the karyotype a missing band from the chromosome.

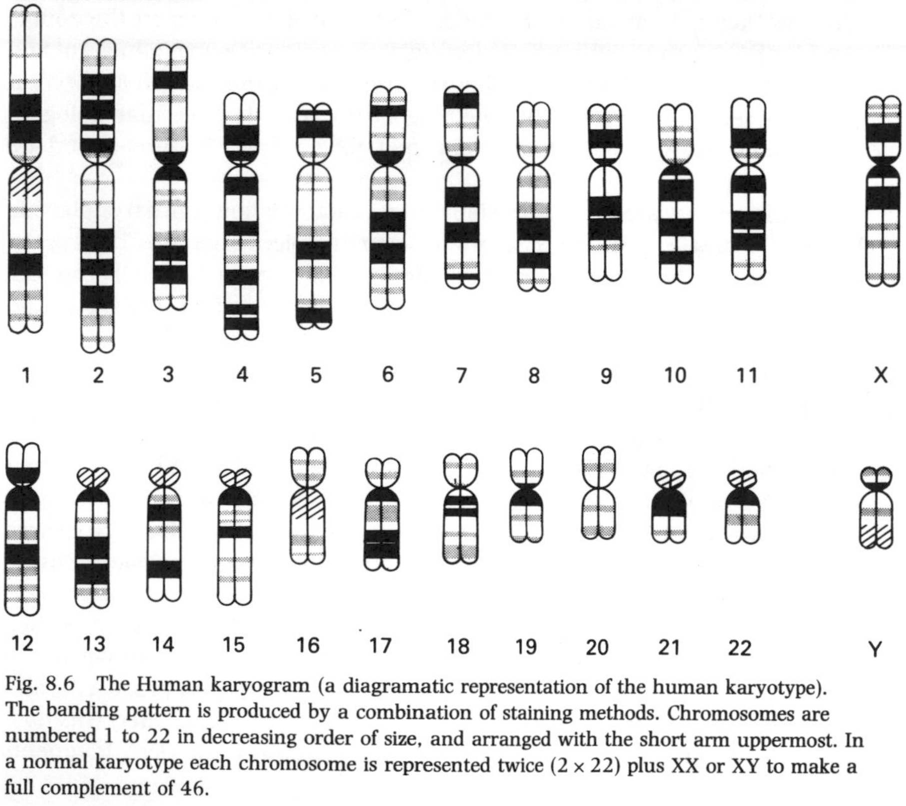

Fig. 8.6   The Human karyogram (a diagramatic representation of the human karyotype). The banding pattern is produced by a combination of staining methods. Chromosomes are numbered 1 to 22 in decreasing order of size, and arranged with the short arm uppermost. In a normal karyotype each chromosome is represented twice (2 × 22) plus XX or XY to make a full complement of 46.

8.7 Gene probes and carrier detection

Today new techniques are being found to locate the alleles of genetic diseases to precise regions of particular chromosomes and to devise genotype tests for the detection of carriers. These methods are available as a direct result of the development of **recombinant DNA technology** which permits human DNA to be cloned and maintained in micro-organisms.

Hybrid cell cultures can be made between mouse cells and human cells. In such cultures numerous different lines of cells may be kept as a living collection of human biochemical activity, for in one line where a particular human chromosome is absent one can demonstrate that an enzyme produced by that chromosome is also missing. This technique, which causes no suffering to human or mouse, has enabled the discovery of many gene locations on particular chromosomes and is also an important tool in cancer research. By another technique large 'libraries' of particular collections of DNA may be cloned in cultures of cells. Kay Davies of Oxford University keeps such a collection of human X chromosomes in a culture of mouse cells.

Another more sophisticated way to locate a gene of interest in a small region of the human genome (the full genetic complement of an individual) is to use a '**genetic probe**'. This technique requires a fragment of DNA, containing all or part of the gene of interest, to be cloned in a bacterium. One way to clone the fragment would be to start from mRNA and to synthesise, *in vitro*, a complementary DNA copy (cDNA) using an enzyme known as **reverse transcriptase**. The next step is to make this DNA double stranded and to insert the double stranded DNA into an appropriate cloning vector using enzymes which cut and join DNA molecules. A myriad of these **restriction enzymes**, which cut DNA at defined base sequences, are now available 'off the shelf' to molecular biologists for this purpose. Enzymes which rejoin the DNA are fewer in number and are known as **ligases**.

Cloning vectors are usually specially made bacteriophage (virus) or plasmid DNA molecules (a plasmid is a naturally occurring piece of circular DNA in the cytoplasm of a bacterium). Using restriction enzymes and DNA ligase, the double stranded DNA copy of the original mRNA is inserted into the vector and introduced into a bacterial cell where it replicates in step with the bacterial chromosome. A **gene library** is a collection of such cloned DNA fragments. Contained within their host bacterium they may be maintained in a culture and stored almost indefinitely at very low temperatues.

Radioactively-labelled **DNA probes** can be used to identify the gene from which the original mRNA was transcribed. Bacterial cultures yield large quantities of vector DNA containing the cloned sequence, which can be excised from the vector using restriction enzymes and can be made radioactive using 'hot' radio-labelled phosphate. The human DNA sample which is to be 'probed', is spread out on a gel by the technique of electrophoresis. The DNA is then transferred (by a method known as 'blotting') on to a nitrocellulose filter which is immersed in a solution containing the radioactive probe. The probe will base-pair with the complementary sequence on the filter and, after washing, autoradiography of the filter will reveal the location of the DNA fragments complementary to the radio-active probe. These fragments are literally one in a million, so it is like finding the proverbial 'needle in a haystack'.

The probing technique can be used to detect carriers of an allele causing a genetic disease if a **restriction fragment length polymorphism** (RFLP) can be found *closely linked* to the allele of interest. This would require study of the DNA in blood samples from a number of close relatives. Human DNA is varied in where particular restriction enzymes will cut it. If some human DNA, from a blood or tissue sample, is cut up with the enzyme that shows a particular fragment length polymorphism then a radio-labelled probe may be used (as described above) to identify the restriction fragments in the region of interest. Individuals or foetuses 'at risk' can be screened to see if they show the form of polymorphism which is linked to the harmful allele. This can be detected in both the homozygous *and* the heterozygous state making the technique useful for the detection of carriers in a high-risk family. This method is used in the foetal diagnosis of sickle-cell anaemia, before any detectable sickle-cell haemoglobin is made (see Chapter 7 Appendix).

A similar technique is used in '**genetic fingerprinting**' (see 11.7). In this case a radio-labelled probe is used to examine the restriction patterns in a highly variable region of the DNA of the genome. In this case, each individual shows a unique pattern which is related in a simple Mendelian fashion to the patterns shown by the individual's parents. This technique is likely to revolutionise forensic science, paternity disputes and immigration cases, where a familial relationship may be proven. The tests, although expensive at present, will soon be mechanised and routine analysis of many samples will eventually be achieved at considerably less cost.

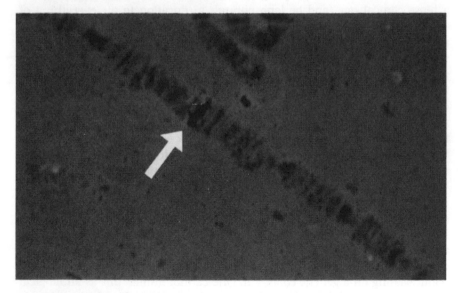

Fig. 8.7 Using a gene probe to find a gene locus. A radioactive DNA probe binds to its complementary DNA on the X chromosome of *Drosophila*. The radioactivity causes silver grains to be deposited in the photographic emulsion on the slide, thus marking the gene locus as a black spot.

Finally, on the research rather than clinical side, a technique using intact metaphase chromosomes, to which the probe attaches directly (*in situ* hybridisation), has enabled great advances to be made with the disease locus detection for such conditions as Huntington's chorea, Duchenne muscular dystrophy, phenylketonuria and cystic fibrosis (see Appendix to Chapter 7).

8.8   Genetic counselling

A genetic counsellor is a trained medical worker who is both a specialist in human heredity and genetic diagnosis and a sensitive and skilled communicator. The aim of the counsellor is to allow people to make well-informed decisions for themselves. Counselling is a participatory activity, not a directive one. The expression 'genetic counselling' was first used to describe this process in 1940 by Sheldon Reed who was tired of the 'naivety, racism, elitism and malice' of the U.S.A. eugenics movement (see Chapter 11).

People will come to the hospital counselling service most probably because of a birth in the family which has involved a genetic defect, or because of some condition that they fear they may have or may transmit. Anybody may use the service, but clearly those that may wish to make use of it, if they know what it offers, include those in risk groups. The list comprises individuals with a known defect, their brothers and sisters and parents or other close relatives, those who marry late, those who have been exposed to large mutagen doses, first cousins contemplating marriage, parents who have had more than one spontaneous abortion, and certain ethnic groups with high gene frequency disorders.

Most of the counselling will follow the necessary diagnostic testing. It should involve both of the parents or prospective parents in a suitable atmosphere, not a tense clinical one, and be long enough in duration to air everything of concern. Skilled counsellors will be able to judge their clients' educational level and psychological state. They will allow them to reach their own decision and will accept that decision if it is at variance with their own view. The counsellor will have to explain all the aspects of the genetical background, and make sure that the 'odds' are understood. For example, a 1 in 4 chance of having a second child with cystic fibrosis could be *thought* to mean that the next three will be normal! Chances sometimes need to be expressed simply and without numbers; there is for example a 'very low' chance of a second child with Down's Syndrome being born to a young mother if she has already had one. Generally a risk greater than 1 in 10 is described as a 'high risk' and less than 1 in 20 as a 'low risk'. The seriousness of the disease is an important consideration; a child with Duchenne muscular dystrophy has currently little chance of long term survival, but parents should be told of the hope that new developments are likely to bring. Here the role of support charities is important. The counsellor will explain the financial, emotional and physical costs of caring for a genetically handicapped person. Parents may have to cope with psychological guilt about passing on a genetic condition. One cannot dismiss this as an illogical fear, for it may well be reinforced in them by the attitudes of others in our society. The tendency to institutionalise the handicapped has been one of the poor features of our past social care; far more open attitudes are found in some 'primitive' societies.

One of the last stages of counselling may well be supplying advice on the options. Some of the choices may not be easy. Counsellors will need to be familiar with the law as it affects the advice they give (Congenital Disability, Civil Liabilities Act 1976) and the grounds given for the choice of a termination of pregnancy (Abortion Act 1967). If a child is still 'planned' by parents in a 'high risk' family there are often other choices. These might include the use of artificial insemination and perhaps in vitro fertilisation (IVF), but more probably there will be favoured options of sterilisation or contraception and the bonus of perhaps adopting an unwanted child. Genetic counselling cannot be done in a hurry and it must be done with sympathy and developed social understanding.

9   Nature and nurture

9.1   Introduction

Any view of human heredity and human diversity that only took genetic influences into account would be sadly lacking. It is all too easy for one to become a classical 'hereditarian' or **genetic determinist**, ascribing the variation between people largely to the genes that they inherit from their parents. The truth may be far from this, for the environment of the individual from conception to adulthood will subtly mould that person in a physical and social way. How would your speech, language, thinking and behaviour differ if you had been brought up in a different country, or in the society of 100 years ago? Because surrounding influences are so important, it is easy to fall into another trap of thinking that differences are all largely the product of environment. The classical 'environmentalist' or **environmental determinist** will see all the outer influences of one's foetal life, birth, childhood and society as being vastly more important than the genetic influence of one's biological parents. Are your academic performance at school and the skill you display on the sports field solely a matter of education and training? Is there a Nobel prize winner and an Olympic athlete potentially present in each of us?

The argument over whether **nature** (heredity) or **nurture** (environment) is the more important in determining human characteristics has been debated for many years. Both sides of the debate still have their protagonists; opinions are often related to people's social and political philosophy and the disciplines of thought in which they have been trained. Although there are no simple answers, it is not difficult to clear up some of the muddled thinking on the subject. Heredity and environment are not two discrete packages of influence. Certainly they are not *qualitatively* the same and cannot be measured in the same way. Both interact at every level to produce the complexity that makes up an individual human being (see section 6.5).

9.2   The environment of the gene

Phenotype or appearance is a product of the **genotype** and the **environment** acting together. This process begins at the level of the cell. The environment of the gene is the sum total of all the factors external to it. Some of these are genetic, for genes interact, and some are environmental. Whether a gene is activated or not may depend on the action of another gene in producing either a suppressor, to switch it off, or an activator, to switch it on. The way that a cell specialises and changes as the body develops depends on the other cells around it. The influence of the internal environment on our development begins therefore from the moment of conception. There is no better example of this than the HY-antigen produced by the Y chromosome. This acts on the gonads to produce testosterone and this hormone then provides the internal environment in which male genital organs differentiate and a male foetus develops.

Phenocopies

Where a phenotype is induced by the external environment to give an appearance normally associated with a genetic effect, the induced phenotype is called a **phenocopy**. Radical alteration of an individual's development may take place in the internal environment of the womb. Two examples of induced phenocopies will illustrate this. **Phocomelia** (literally 'seal limb') is a condition in which a child is born with extremely short arms and legs that look, at birth, like seal's flippers. Although there is certainly a genetic basis for this rare condition, it became alarmingly common in the 1960s in Europe and the U.S.A. when the drug **thalidomide** was prescribed, as a sedative, to pregnant mothers. Some characteristics of this drug, operating in the embryonic environment, prevented normal limb development. Fortunately those children who suffered this physical handicap are genetically quite normal and many of them are now the parents of normal offspring. Deafness may either be genetically caused or it may be induced in a child through *Rubella* infection of the mother at an early stage of her pregnancy. It may not be easy to distinguish the causes of such deafness later in life.

9.3 Twin studies

Francis Galton (1822–1911), a first cousin of Charles Darwin, was one of the first people to generate scientific interest in the 'nature *versus* nurture' debate. He was largely responsible for the growth of biometrical science and being a very able mathematician made careful measurements against which his theories were tested. Today we can view him as a rather fanatical hereditarian. It is to his credit that he was one of the first to recognise that identical twins were genetically the same. He perceived the value that they presented to precise science for measuring the different parts played by heredity and environment. Although 'twin studies' have come into some disrepute through the alleged falsifications of Cyril Burt (see 9.5) they are worthy of some serious consideration.

Monozygotic twins (MZ twins) are 'identical' in that they have the same genotype, resulting from the one zygote making an early cleavage and the two halves separating to form the two embryos. **Dizygotic twins** (DZ twins), which are several times more common, differ in being the result of two fertilisations. These are only as genetically alike as any two brothers or sisters born to the same parents. DZ twins may be of the same or different sexes but MZ twins must, of course, be the same. Table 9.1 shows the genetic relatedness of different individuals in a family. It will be seen that MZ twins have a closeness that is quite unusual. They are effectively two members of a **clone**. All their cells are genetically identical and consequently organ transplants between them are easily possible.

Table 9.1 The genetic relatedness of members of a family

Relationship	The minimum proportion of alleles held in common	
Monozygotic twin to twin	1.0	all
Dizygotic twin to twin	0.5	$\frac{1}{2}$
Brother/sister to brother/sister	0.5	$\frac{1}{2}$
Mother/father to child	0.5	$\frac{1}{2}$
Grandparent to grandchild	0.25	$\frac{1}{4}$
Uncle/aunt to nephew/niece	0.25	$\frac{1}{4}$
First cousin to first cousin	0.125	$\frac{1}{8}$

If MZ twins are genetically the same they will only differ in *any* respect because of environmental influences. DZ twins differ in phenotype for both environmental and genetic reasons. Because the MZ twins differ, in the single variable, and the DZ twins because of two variables, the extent of MZ and DZ differences, it is argued, will show up the extent of the genetic difference on its own. Monozygotic twins are quite obviously more alike, but for this comparison to be legitimate we would have to assume from the outset that the environment influencing both of these different sorts of twins, in the way that they are treated as they grow up, is the same. We shall return to this point later for it is not a valid assumption.

For any measurable trait, or characteristic, twins may be said to be **concordant** if they both measure the same for it, and **discordant** if for the particular trait they both measure differently. (Clearly the size of the measuring units affects a decision on concordance and discordance, for example two people of the same height are not *exactly* the same height, but may be so to the nearest centimetre.) If a number of easily recorded features (see Table 9.2) are considered for a sample of over 50 twins of each type, for no characteristic are the DZ twins more concordant than the MZ twins. Given that both sets of twins are reared together, the difference in concordance will reflect the relative importance of the hereditary component. The most obvious way that MZ twins are concordant is by sex (which we know is entirely genetic in origin) so to make comparisons, that discriminate heredity from environment better, only DZ same-sex pairs are compared with the MZ twins. Table 9.2 shows the rank ordering of some physical features and disease susceptibilities, from the greatest to the least heritability component.

Although this simple examination of the MZ–DZ concordance differences is rather subjective and qualitative and may lack statistical rigour, clearly some physical features such as eye colour have a high hereditary component. Similarly there is little doubt that susceptibility to certain diseases such as rickets and some psychological conditions, although environmentally triggered, do have some hereditary element in them. Even for a highly contagious disease like measles there may be some genetic component affecting susceptibility.

Table 9.2 Concordance studies in twins

Physical feature	Concordance % MZ	DZ	MZ-DZ difference
Eye colour	99	28	71
Fingerprint ridge count	95	49	46
Height	95	52	43
Mean blood pressure	63	36	27
Mean pulse rate	63	36	27
Diseases	Concordance % MZ	DZ	MZ-DZ difference
Rickets	88	22	66
Manic depressive psychosis	73	16	57
Diabetes mellitus	70	23	47
Schizophrenia	52	12	40
Tuberculosis	53	22	31
Scarlet fever	65	47	18
Measles	95	87	8

Source: Jenkins, 1983

Taking this kind of data seriously does presume that MZ and DZ twins have the same sort of environmental upbringing. Is this true? Identical (MZ) twins tend to spend more time in each other's company; their parents like to emphasise their similarity in the way that they dress them, and more often prefer them to go out together. In the U.S.A. they even have 'twin conventions' with prizes for the most similar twins! Such effects will *increase* concordance, for example picking up the same germ and having a more similar social environment. Certainly MZ twins tend to hold together for much longer in life. If one MZ twin is diagnosed as schizophrenic is it not more likely that the other will be so classified if he or she shows similar schizoid behaviour? One solution to this criticism is to compare MZ twins reared together with MZ twins reared apart from each other, after separation at birth. This is fine as an experimental method, but examples of the latter kind are hard to find and certainly not easily available in the numbers that are needed for convincing proof. (So hard are they to find that the famous psychometrician, Sir Cyril Burt, is accused of having invented data on such twins, to strengthen his belief that intelligence was largely genetically inherited!)

9.4   A measure of heritability

The extent to which the variation we observe in a population is due to genetic differences or to environmental ones can only be properly quantified by a statistical method known as **analysis of variance**. If we consider a trait like height and analyse the variation in recorded heights between a large number of MZ twins reared together and between DZ twins reared together, we may calculate the variance (the square of the standard deviation). If a factor has a high 'heritability' the variance for the MZ twins will be very small when compared with the DZ twins. If, by contrast, something is entirely environmentally determined, the variance of DZ and MZ will be larger and more closely similar between the two types of twins. The size of the variance depends on the size of the units used but will not greatly affect the measure of 'heritability' expressed as:

$$H = \frac{V_{DZ} - V_{MZ}}{V_{DZ}}$$

Where

V_{DZ} = variance between dizygotic twin samples
V_{MZ} = variance between monozygotic twin samples
H   = a measure of heritability (between 1 and 0)

This method is illustrated by Table 9.3 which is based on a study of the height differences between ten twins (a) to (j) of each sort.

Table 9.3 Analysis of variance for measuring heritability of height in 10 twin pairs

Sample	*Difference in height (cm) of MZ twins reared together*	*Difference in height (cm) of DZ twins reared together*
a	1	6
b	2	3
c	2	14
d	6	1
e	1	6
f	2	10
g	4	8
h	0	12
i	3	4
j	1	7
Mean x	2.2	7.1
Standard deviation SD	1.75	4.04
Variance (SD)²	3.07	16.32

$$H = \frac{V_{DZ} - V_{MZ}}{V_{DZ}} = \frac{16.32 - 3.07}{16.32} = 0.81$$

Source: Jenkins, 1983

The result may be considered to imply that genes contribute 81% of height and by inference 19% comes from the environmental factors (such as nutrition). We may ridicule the precision of such figures and the simplistic interpretation of two discrete components, but those who approach problems in such mathematical ways confidently quote such numbers as if they were 'hard science', not open to doubt. Whatever one's feeling they do reflect *some* reality, but deserve to be treated with scientific rigour to be at all valuable. (See Table 9.4.)

Table 9.4 'Heritability' of some commonly measured human attributes as calculated by analysis of variance

Attribute	H
Height	0.81
Weight	0.78
Brain size	0.75
Intelligence Quotient (IQ)	0.67
Verbal aptitude	0.68
Spelling ability	0.53
Arithmetic aptitude	0.12

In some important areas of concern, such as the educational and social development of children, and in certain psychological disorders, the influences of nature and nurture are clearly of legitimate interest to us all. It is therefore worth looking at three cases more fully: IQ, schizophrenia and gender differences.

9.5   Intelligence Quotient (IQ)

Estimates of the heritability of intelligence, as measured by the analysis of variance from IQ test scores, vary between 0.34 and 0.85, with most recent estimates being less than 0.5. Such variations in the estimate, themselves the means of even wider variations, clearly indicate that either different people are measuring different things, or 'intelligence' is not a readily defined entity at all! Most scientists today feel very uncomfortable about the seeming precision of these measures (Rose, Kamin and Lewontin, 1984). It is, after all, a truism that all that IQ measures is an ability to perform in an intelligence test! Much significance has been traditionally attached to IQ and it is commonly (and erroneously) asserted to be a precise measure of innate, or inherited, ability.

The first standard 'intelligence test' was devised by Alfred Binet in Paris in 1905 to define the educational level of children relative to their age. By defining a range of tasks that an average ten-year-old could perform, Binet introduced the concept of **mental age**. His test's original purpose was to distinguish children in need of remedial help with their schooling. Binet readily recognised that children from more privileged classes of society performed better on his tests, for their age, than the average Parisian child. In the ten years following 1905, Binet's test was developed and modified in the U.S.A. to make it more quantitative, using a linear scale and the assumption that the range of intelligence must fit a **normal distribution curve**, with the average intelligence in the middle of the 'camel's hump' distribution. The concept of mental and chronological age comparison thus gave way to a distribution of test attainment on a quotient scale of 100 points for the average performer. Terman, who introduced the **Stanford-Binet Intelligence Quotient Test** in 1916, was a eugenicist (see Chapter 11) with firm and fixed ideas about the major component of intelligence being heritable, and the most intelligent people being of Northern European descent and of the upper social class. Thus the original intention of the test was changed; the IQ test evolved into a diagnostic tool in education whereby children were selected for appropriate courses on what was presumed to be their inborn level of ability. The tests themselves have in the past been very culturally biased and one should always view them with caution. Sir Cyril Burt (1883–1971), Britain's pioneer of IQ testing, must take some responsibility for the disrepute of the IQ test and indeed all testing of this type. (See S. J. Gould, 1981.) As a young man Burt came under the strong influence of Galton and Karl Pearson, the two hereditarian pioneers of the British eugenics movement. Although he did much to develop such work as twin studies, he was so convinced of the hereditarian position that he seems to have fabricated data to support the hereditarian view. It is interesting to note that Burt's work was instrumental in establishing the selective 11 + examination that was the basis of the division of English secondary education into the grammar and secondary modern streams, before the introduction of comprehensive schooling. The notion still prevails in much of British schooling that 'failure' in an exam is synonymous with some heritable deficiency of innate intelligence. Today the conviction that intelligence is very largely inherited is upheld by many psychometricians such as Hans J. Eysenk and Arthur R. Jensen. Both of these still maintain that the slightly lower mean performance of 'blacks' on their IQ tests measures a *real* genetical difference between races that is independent of cultural bias. Many geneticists are sceptical of their certainty. It is easy to include a cultural bias in a test and just because the test is quantitative it is thought to be measuring a reality. One of the dangers in science is to assume that measured things are 'real' and (subconsciously, therefore), immeasurable ones just don't exist!

Two recent studies of the heritability of intelligence have looked at children adopted into families where they have been brought up with the natural (biological) children of the parents. Fig. 9.1 shows the influences that operate in this case.

Fig. 9.1. Separation of the effects of genes and environment by comparing adopted and biological children in the same family.

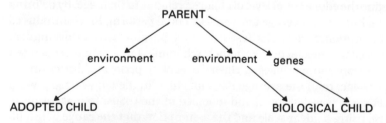

Source: Scarr and Weinbert, 1977 and Horn, 1979.

The advantage of this type of study is that such families are far easier to find than ones with MZ twins separated at birth. Much larger samples are possible. The IQ of the mother and her biological child are compared with the IQ of the mother and her adopted child. The results of such studies have been that children reared by the same mother resemble her in IQ to the same degree, whether they share her genes or not. When the adoptive father is included in the studies there is not such a clear correlation. Importantly, some adoptive studies of this type have been done across race and class boundaries. What is interesting is that 'black' children adopted into 'white' homes in the U.S.A. show higher IQs than the mean for 'whites' of that age group in the general population. The adoptive homes are probably 'good homes' of higher than average social class. Two other studies, in which the amount of 'white' genetic ancestry possessed by 'black' children is compared with their IQ scores, have found no correlation. This would point to race itself being irrelevant in the variation of intelligence that exists. Certainly 'blacks' in the U.S.A. have mean IQ scores below those of 'whites', but their socio-economic status is much lower and so is the mean quality of their schooling. On the other hand, if IQ does measure anything meaningful and if inter-racial differences in IQ scores do exist, all other things being equal, it would be wrong to dismiss them or pretend that they do not exist. The real lesson is that individuals vary very widely within a group and that the human intellect has enormous diversity and dimensions not readily tested. Without a stimulating childhood environment, intelligence cannot develop on its own.

9.6   Schizophrenia

Schizophrenia is a psychological illness with both a genetic and environmental component. Those with a tendency towards schizophrenia are believed to be above 1% of the population. These people may experience delusions and hallucinations and may manifest bizarre responses to normal stimuli. They often appear to have a loss of interest in life's normal activities and pleasures. The condition is unrelated to intelligence level and most schizophrenics are not a danger to themselves or other people.

Schizophrenia seems to be varied in its degree and manifestations, either because it is due to a dominant allele governed by other genes, or because it is polygenically produced. The fact that only 52% of monozygotic twins are concordant shows how variable its development is, even when they live so close together. If it was just a matter of genes alone this figure would be close to 100%. The concordance for schizophrenia in DZ twins, and ordinary brothers and sisters, is 12%. This is very much lower. Environmental influences of some kind are therefore involved in precipitating some of the severest conditions; what the influences are we do not know. What genes there are that are involved in schizophrenia are not without benefit. Individuals with mild schizoid behaviour are amongst the most creative and artistic members of the human community.

9.7   Gender roles and sexual discrimination

Another area where an understanding of the roles of heredity and environment is very important is that of sexual differences. What do we mean by 'equality' between men and women? The last few million years of evolution have left their mark on us as hominid hunter-gatherers. Women have obvious adaptations to more sedentary child rearing whilst men have obvious adaptations to more physical exertion. This does not mean that men are incapable of rearing children nor women of working physically harder than men. We must expect to find both biological adaptations and variation in natural abilities.

If we accept the norm referenced testing (looking to see if each thing is above or below average) of the type employed in the IQ test, then *on average* women may be found to have a stronger immune system, greater sensitivity to taste, higher tolerance of pain, more acute hearing in the high frequency range, a better sense of touch and a greater awareness of a wide variety of sensory stimuli. They are supposedly more caring, more moved by emotion, more sensitive to body language, better at remembering faces and more articulate in speech. On the other hand, *on average* men are physically stronger, faster running, more aggressive, more competitive, better at analysis, more single-minded in pursuit of goals, more able to shut out distractions, and have better mathematical and spatial abilities.

What is concealed by these statements is that for these variables there is an enormous overlap between the distributions for men and women about a *slightly different* mean. What cannot be stated enough is that for any of these features there is far more range of difference between *all* men and between *all* women than there is between the means for the two sexes.

So far as genetics goes, many of these traits must be polygenic and in theory are unlikely to be linked to sex chromosomes. What then makes the difference? Many of the characteristics, like aggression or mood, may be influenced by the hormonal environment of the body. Very probably the internal environment provided by the sex hormones makes a lot of difference. Other gender differences may have an overwhelmingly cultural component, and disentangling heredity from environment in this case is well nigh impossible. Certainly the social environment we live in, and the parental expection of what is appropriate behaviour for young boys or girls, influences the subsequent development of our own attitudes and skills. Do you think that your parents were 'sexist' in the toys they gave you to play with? How many boys are presented with a doll to nurse and how many girls are given mechanical models to build, or toy guns? Any one man or woman has a constellation of the characteristics that emanate from heredity and environment. There is no absolute biological equality between males and females, but there is so much diversity that it makes no sense to restrict particular jobs or activities to either men or women exclusively. It is right to insist on 'equality of opportunity' between the sexes, but that does not mean that there is 'biological equality'.

9.8 The good cake analogy

One analogy that serves well in the nature–nurture debate is 'the good cake'. Imagine, for a moment, the very best cake you can think of. Suppose that the raw ingredients are the 'hereditary' components and that the preparation, baking decoration and presentation are the 'environment' to which these ingredients are subjected during the production of 'the good cake'. Can you say what percentage of the *goodness* of the cake is due to the ingredients and what is due to the way that it was produced? Genes and environment interact from the very start of development. Both are intricately essential but because they are *qualitatively* so different in kind there may be little point in trying to discover which is more important.

10 Genetics and race

10.1 Introduction

The human species has a quite remarkable diversity that extends beyond the variations within local populations. The term 'race' may have an emotive tone but it is used with some precision by biologists as a term describing the geographical variety within an animal or plant species. Yet 'race' is also an expression that describes a social concept of differences between peoples, for human groups commonly recognise differences amongst themselves not only in physical features, but also in ethnic origin and cultural allegiance. The biological concept of race and the social concept of race are thus rather different and no useful purpose is served by tangling them together. For the social biologist, who looks at the biology of the human species, we therefore have two questions about race. Firstly what is the biological diversity of adaptations that exists in our species and secondly how do different ethnic groups behave towards each other? In looking at these separate questions we need to be aware of two extreme viewpoints. On the one hand one needs to avoid the 'racist' assumption that one's own particular ethnic group is biologically and culturally superior to others and that existing 'races' are somehow fixed and should not be mixed by intermarriage. On the other hand we should not pretend that all people are biologically identical and that human differences are non-existent.

10.2 Biological race

The biological concepts of species and race need to be clear. A **species** is

'a group of interbreeding or potentially interbreeding natural populations which are reproductively isolated from other such groups'   (Mayr, 1963).

Reproductive incompatability prevents one species from mixing with another. Today, at least, there is only one species of human, *Homo sapiens*. Because they interbreed, members of a species share a common **gene pool**. However widely spread they may be, all human populations are interfertile; no barriers of reduced fertility are to be found between races, indeed if anything the opposite is true since there is genetic **hybrid vigour** in outbred populations.

In evolution, once a population is reproductively isolated from another, that is for whatever reason the two cannot mate successfully to produce fertile offspring, there is no going back to reproductive compatability. Each population is likely to diverge until the two must be regarded as separate species. Such **speciation** has produced the branching tree of all evolved forms.

What then is a race? The biological concept of a **race** is of

'a subdivision of a species formed by a group of individuals sharing common biological characteristics that distinguish them from other such groups'   (Mayr).

Race is the smallest classificatory group defined in taxonomy. It always implies a geographical variation or ecological type. The biological characteristics that distinguish races, of any species, are therefore environmentally adaptive ones. Anthropologists, who study the human races, would add to this taxomonic and geographical definition of a race the sharing of common cultural characteristics. Cultures change, as do living organisms, and hence isolated races are likely to evolve different languages and cultures from their biologically slightly different neighbours. Where two geographical races meet individuals are found in a gradation between the two differently adapted types. Such interfertility is, from a biological standpoint, a sure sign of ancestral links joining all human communities together. All members of the sub-species of modern humans, *Homo sapiens sapiens*, share ancestral links that diverged perhaps less than a quarter of a million years ago.

10.3   Human races

One may be asked, when filling in a form, for one's 'race'. Biologically there is no set way in which one can answer this question. The terms 'white' and 'black' are not races, but colours, and as such are social expressions. Nowadays one is more commonly asked for one's 'ethnic origin'. This question is not a matter of biology but of personal and cultural identity (see 10.7), to which it should be easier to give a satisfactory answer.

Today biologists recognise that the genetic variation that is found within different ethnic groups is much greater than that to be found between them. This is another way of saying that race is not highly significant in a genetic sense. However, we must recognise that 'race', as perceived objectively by anthropologists and population geneticists, is an expression of human diversity and adaptation to different environments. For this reason we must look at attempts to distinguish races. The anthropologist, C. S. Coon (1962) recognised five racial groupings of a zoogeographical kind. These are the **Capoids**, the (cape) bushmen of southern Africa; the **Negroids**, the other black people south of the Sahara; the **Caucasoids**, the Mediterranean, European and West Asiatic people; the **Mongoloids**, the East Asiatic people and indigenous American Indians of North and South America; and the **Australoids**, the people of New Guinea and the aboriginal people of Australia. Coon's classification is valuable in distinguishing two ancient hunter-gatherer peoples (bushmen and Australian aborigines) and in defining the negroid group to its African origin. It helps less well to sort out the diversity of people in Euro-Asia other than to recognise that there are Western forms typified by the people of the Caucasus mountains, between the Black Sea and the Caspian sea in the Southern U.S.S.R., and eastern forms typified by the Mongols of the far Eastern U.S.S.R. Coon recognises that the Mongoloids have spread from Eastern Asia across into the Americas, down to Tierra del Fuego and seawards from the Americas to Polynesia.

Because there are so many geographical regions of overlap, others have made finer subdivisions. T. Dobzhansky (1962), taking a population genetics viewpoint, recognised *fourteen* races each with a more defined gene pool. This classification, for example, puts the Maoris of New Zealand as a definitely distinct race from the very distant Eskimos of North America. S. M. Garn (1961) has made further subdivisions and recognises 32 human races, each uniting many 'tribes' of clearly biological and cultural descent. Even with Garn's groupings the boundaries are blurred, for any knowledge of history teaches us that different ethnic groups have risen and fallen over the centuries according to their relative success in competing with and assimilating their neighbours. Clearly therefore there are many gene pools of our species but the geographical barriers between them have never been insuperable from the Palaeolithic to the present day. The diversity of human physical adaptations, that largely define ideas of 'race', tell us more about the wealth of our species' history than anything else. S. J. Gould (1980) feels that it no longer makes good *biological* sense to talk of human races as discrete units. It is instructive to read his essay 'Why we should not name races'. (See Bibliography.) Many contemporary biologists would agree with him.

10.4 Natural selection and human diversity

Few biologists doubt that evolution has occurred very largely by natural selection. The expression 'survival of the fittest' summarises the working of the selection process, but who are the fittest? Neo-Darwinian theory sees **fitness** as meaning that the carriers of a particular genetic variation leave more surviving offspring than the possessors of other such variations. This means that the individuals that survive better to produce more offspring, as judged over a lifetime, are 'fitter'. Darwinian fitness is therefore a product of **viability** and **reproductive success**. Such individuals are so well adapted to their environment that they do better than other individuals. Natural selection acting on a population in one place results in **adaptation**. Many human racial characteristics, such as colour, are plainly valuable environmental adaptations. (See 10.5.)

One of the key characteristics of a successful organism is a capacity to adapt to a changing environment. Thus selection also operates in favour of **adaptability**, which is the capacity to adapt. Although the different human races may have various environmental and behavioural adaptations, individuals may have both an amazing environmental adaptability, such as to the global extremes of hot and cold, as well as behavioural adaptability, such as in the learning of different languages.

Charles Darwin was of the strong opinion that much selection of human racial characteristics was due to the mating preferences exercised by marriage partners' perceptions of beauty and handsomeness. He wrote in 1871,

'I conclude that of all the causes which have led to the differences in external appearances between the races of man sexual selection has been by far the most efficient'.

Mate selection, by males of females and by females of males, is as complex a subject as it is at present speculative. If the possessors of particular features have a higher fitness, and so produce more offspirng, that physical form will come to be more prevalent. A society with social hierarchy and polygamy will certainly produce selective effects of this sort. It is possible that the evolution of skin colour, the absence of body hair, the presence of abundant distinctive head hair, red or fleshy lips or sheer tallness are all examples where mate choice may have been influenced by perceived attractiveness. Perceptions of male and female beauty differ in different cultures and at different times. For example, portraits and literature in Western Europe show that in past centuries plumpness and a very white skin in a woman were a sign of beauty. Today's fashion models are contrastingly skinny and suntanned. A black skin is beautiful in many cultures and is assertively so in the 'black consciouness' movement amongst people of African origin today. Gender signals and cultural fashion are still powerful in their influence upon us. However, for sexual selection to be effective it will need to operate in a consistent direction for thousands of years to cause any major change.

Natural selection is not all of one type in its influence upon a population. **Stabilising selection** involves selection for a particular characteristic against the extremes. This is well shown by Fig. 10.1 which shows stabilising selection for birth weight in the U.K. Children born overweight or underweight are at a disadvantage. Stabilising selection will give a local population a more centred characteristic for a particular variation. However, what is normal in one environment may be abnormal in another and hence different races may be normalised or centred on different points in the total human range of variation. **Directional selection** is where the selective pressure favours one extreme of the natural variation so resulting in quite fast evolutionary change in the population as a whole. Amongst the Watutsi of Eastern Africa social status is related to

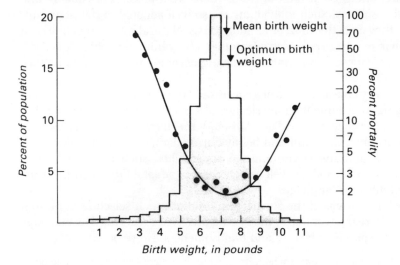

Fig. 10.1 Stabilising selection for human birth weight. The histogram shows the proportions of babies that fall into each birth weight class. The curve plots the mortality around the time of birth against the birth weight. The optimum birth weight for survival is very close to the mean birth weight. (US data, after Cavalli-Sforza, 1976)

height. It is not surprising therefore that these are the tallest people in the continent of Africa. **Diversifying selection** is where there is selection for a wide variety of different attributes, the extremes of which, if selected, will confer greater fitness on individuals within the population as a whole.

Armed with a knowledge of the variety of selection effects and an awareness of environmental adaptations, the biologist can *speculate* about some diverse human features! The shape and form of the human nose is one such example. The early hominid nose was in all probability ape-like, being very short and flattened and with widely spread and broad nostrils. Such a nose would be adapted to warm and humid conditions. Amongst hot desert-adapted humans, such as the Jews and Arabs, and cold desert-adapted people, such as the Inuit Indians (Eskimos), the nasal passages are much narrowed and nostrils closer together. Such desert adaptation, which is paralleled by many other mammalian species, has survival value in conserving moisture or aiding thermoregulation. The relatively long noses of many Europeans are believed to be a desert-adapted feature from their evolutionary past. Amongst the Mongoloid people, whose ancestors lived through many thousands of years of ice-ages, physically flatter noses that do not protrude from the face may have had survival value in reducing wind chill to an exposed face. (In Siberia the mean annual temperature is $-18°C$ and the January mean $-53°C$.) Some East Asians of China and Japan have a cultural preference for short noses, finding some European noses oddly large and long. In Britain a long narrow nose has been considered 'aristocratic' and the same is true amonst the Hindus. Depending on your cultural upbringing, a 'handsome' or 'beautiful' nose may be large or small, short or long, snub or pointed, broad or narrow. Although these observations are very speculative they serve to illustrate the diversity of one human facial feature that has been subject to a variety of types of selection.

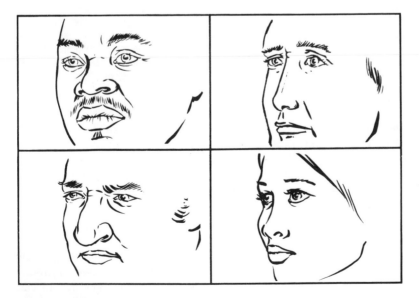

Fig. 10.2   A diversity of human noses

10.5   Environmental adaptations of humans

Some human features show very clear **climatic adaptations**. Amongst mammals in general there is a decrease in the length of body extremities (such as ears and limbs) as one approaches colder climatic regions (Allen's Rule). Heat is thus conserved in the cold and more readily lost in the heat. A person's sitting height divided by their standing height gives a useful index of the relative shortness of their lower limbs. Many Africans, Australian aborigines and some Europeans have long narrow limbs with a sitting height/standing height ratio of between 0.45 and 0.50. Amongst the more cold adapted Eastern Asians and North American Indians the ratio is between 0.53 and 0.54. (See Fig. 10.3.)

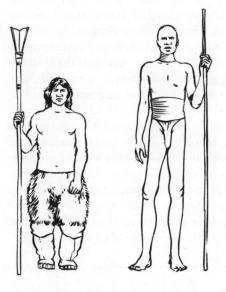

Fig. 10.3   Limb proportions of different humans. Nilotic people of Eastern Africa are tall and slender with relatively long legs. This may be regarded as an adaptation to heat loss in a hot environment. The Inuit (Eskimo) of Greenland is short-limbed, arguably an adaptation to conserving heat in a cold environment.

Amongst all of Coon's classical races there are very tall and very short sub-groups. A diminutive stature similar to the pygmy people of central Africa is found amongst unrelated equatorial forest dwellers, such as the original inhabitants of the Andaman Islands, Malaya, The Philippines and Papua New Guinea. Short height may be a forest dwelling adaptation of these hunter-gatherers, many of whose quarry species are much smaller than their savannah counterparts. African pygmy children at the age of three are of average height, when compared to other races, but by the end of adolescence they are the shortest of any ethnic group in the world. This adaptation has a simple genetic basis. Pygmies produce markedly less of a pituitary growth hormone (IGF1) at adolescence, growing only 20 cm in height when the average African child would grow 32 cm.

Skin **pigmentation** is the most strikingly obvious racial difference. The basal layer of the epidermis continuously produces new cells to replace those worn away from the outer dead layers of skin. All humans have equal numbers of **melanocytes** (cells) in this basal layer but the number and size of the **melanosomes** (black melanin bodies) that they each contain varies between the races. Exposure to the shorter wavelength radiations of sunlight stimulates the melanocytes and a protective tan is produced. The ultraviolet radiation can damage the cells quite seriously and it is well known that excessive radiation may cause skin cancer in the basal skin layer. The most serious skin cancer is called melanoma, and is fatal in about 50% of cases. Skin cancers are most frequent in white populations living in areas of intense sunlight. It is not surprising therefore that the distribution of *indigenous* black people follows the tropical belt from Africa, through southern Asia, to the appropriately named Melanesian region of Papua New Guinea. Within a large land mass of ancient habitation such as Africa or the Indian sub-continent there is a lightening of skin pigmentation from the equator towards the more temperate latitudes. Given the African origin of our species it is important to ask why white people are so unpigmented. The answer here seems to lie in the well-known benefit of UV light in the synthesis of vitamin D. It may well be that the earliest Nordic European people benefitted so much from this critical nutritional factor, particularly if they were clad in clothes, that becoming less pigmented was of selective importance. Whatever the real answer it is probably environmental adaptation that gave the first steps towards being whiter or blacker. Culture-based sexual selection for being beautifully black or beautifully white may have made a difference too.

Human hair is very varied. All Mongoloid people and Negroid peoples have entirely black hair, but Europeans, bushmen and aborigines have some blonde and red hair alleles. As it is in children that these non-black hair colours appear most markedly, selection of blonde hair may be by the retention of a childhood genetic character. If circular in cross-section, hair will be straight; if more elliptical in section, it will be wavy; if flattened from side to side, it coils up naturally into a tight spiral form. Tightly coiled Negroid hair provides the best protection for the brain from the adverse heating effects of the sun.

10.6   Human diversity and disease

One of the striking racial differences between peoples are the variations in blood groups and blood pigments that they possess. Certainly some of these may be associated with resistance to particular diseases.

The ABO system is very ancient, being found in the great apes as well as man. It is an example of a **balanced polymorphism** where selection is diversifying. Each of the alleles A, B and O, must possess some situational advantage to be maintained. Although the O group is the most common there are striking differences in the individual allelic frequencies in different areas of the world. Is this just a matter of chance, or genetic drift, or does it reflect adaptation to the environment? Group A is most frequent amongst Australian aborigines and the Lapps of northern Europe, whilst group B is highest in China and Central Asia. Both alleles A and B are rare amongst American Indians. There appears to be a fitness advantage to be gained from being heterozygous but it is unclear why this should be. Vogel (1960) has suggested that group O is frequent only in areas where there has been no major epidemic of plague (*Pasturella pestis*) and that group A is most frequent in areas where smallpox was endemic. In Central America the arrival of smallpox with the 16th and 17th century Europeans did more to destroy the indigenous native civilisations than the strength of the invading armies.

Malaria, due to *Plasmodium falciparum* or *P. vivax*, is still the most widespread killer in the humid tropics and sub-tropics. The malaria parasite invades red blood cells and the host is only naturally protected if these cells can resist invasion. There is much evidence to show that not only sickle cell anaemia but also the genetic disorders of thalassaemia and glucose–6–phosphate dehydrogenase (G6PD) deficiency are also related to malaria.

Sickle cell anaemia (see Appendix to Chapter 7) manifests itself today as a major genetic handicap for people of recent African origin. The condition is named after the sickle-shaped cells that are found in the blood smear test. Most black people have only haemoglobin A (coded for by the allele HbA) but sickle cell sufferers have only the haemoglobin S (coded for by the allele HbS). For these people a lowering of oxygen tension can bring on a sickling crisis in which the cells are distorted and easily broken to bring on anaemia. The heterozygotes (genotypically HbA/HbS) show virtually no sign of the sickling, or of the anaemia, but are carriers of the disease. The geographical distribution of endemic malaria and sickle cell disorders are given in Fig. 10.4.

Sickle cell disorder is a clear cut example of an environmental adaptation under a **balancing selection**, for the heterozygotes have a Darwinian fitness greater than both the normal homozygote HbA/HbA individuals and the

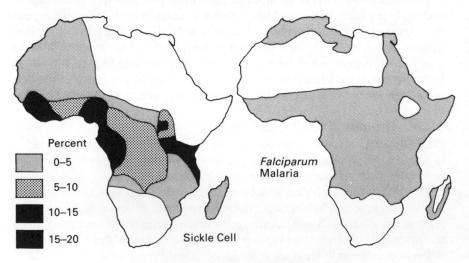

Fig. 10.4   The distributions of malaria and sickle cell disorder in Africa

homozygote HbS/HbS individuals. It seems that when heterozygotes are infected with malaria by a mosquito the malaria parasites are unable to survive in the red blood cells. Whilst the homozygotes HbA/HbA suffer from malaria the heterozygotes are relatively immune. Evidence for this link with malaria comes from the geographical correlation, direct experimentation (Allison, 1954), and the observation of higher parasite counts in HbA/HbA malarial patients and their differentially higher mortality rate amongst infected patients in hospitals. The greatest gene frequency of HbS recorded is 16% and it occurs where people commonly experience one or more bites per night by disease-carrying female *Anopheles* mosquitos! All people of recent African or West Indian ancestry in Britain should know that they have approximately a one in fifteen chance of being carriers.

Other examples of genetic variation between racial groups that are associated with malarial disease incidence are those of β thalassaemia and G6PD deficiency, in Southern Italy, Sardinia and Greece. *Plasmodium falciparum* used to be endemic in Sardinia but is now no longer found. Village records for malaria incidence were kept (1929–1938) before eradication of the disease. More recent gene frequency studies (1961), for the two blood disorders have been carried out in the very same villages from which the old records come. These show that higher gene frequencies are present among the direct descendants of those who once commonly had malaria a generation before. People of Italian, Greek and Turkish ancestry in Britain, therefore, are more likely to be carriers of β thalassaemia than other Europeans.

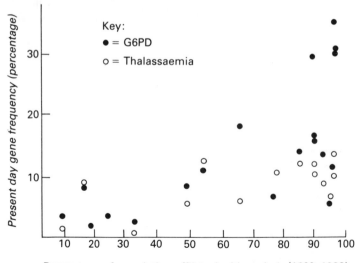

Fig. 10.5 The correlation between malaria and the gene frequencies for the thalassaemias and G6PD deficiency in Sardinian villages (see text). (After Cavalli-Sforza, 1963)

It is easy to fall into the trap of thinking that such genetic conditions that have evolved in response to an environmental factor, like malaria, are oddities amongst the immigrant African or Mediterranean people in Britain. We should remember that such immigrants are much *less* likely to suffer from phenylketonuria and cystic fibrosis, for these are markedly more common in Northern Europe. CF (see Appendix to Chapter 7) is a much bigger burden on the nation's health than sickle cell disease. With a biological view of human genetic adaptation, we should be asking why cystic fibrosis is carried by one white person in twenty-five? Is there a possible heterozygote advantage?

10.7 The origins of modern races

Studies of blood proteins and **mitochondrial DNA** provide a molecular clock from which deductions can be made about the way in which the major races may have originated from a common African origin. Such evidence backs up anatomical and archaeological theories about the recent human past. Fig. 10.6 shows a cladogram based on variations in blood group gene frequencies. These data suggest an early division between the Asian and New World forms of *Homo sapiens* on the one hand and the African and European forms on the other.

More recent investigations are based on the studies of mitochondrial DNA. Mitochondria are organelles found at some stage in all cells. They are unusual in being the only animal cell organelle, apart from the nucleus, to contain DNA, possessing a very short length of 16 kilobases. Mitochondria are maternally and cytoplasmically inherited; all of the mitochondria we possess come from the zygote that gave rise to each of us, the cytoplasm of which is all our

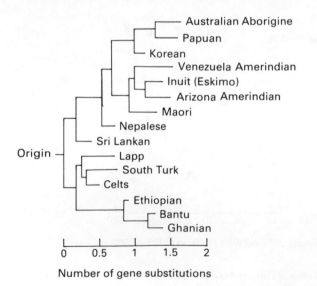

Fig. 10.6 A cladogram of racial kinship based on blood group gene frequencies. This genetic data strongly supports evidence from physical anthropology on human evolution.

mother's. Mitochondria are therefore inherited in the female line; from one's mother, from her mother (one's maternal grandmother) and from her mother before that and so on. Mutation in this mitochondrial DNA (mDNA) is random and accumulating. A knowledge of the extent of the sequence change therefore acts as a record of the duration of the separation of two maternal lines. From such DNA sequencing studies it has recently been shown that the greatest diversity of such mDNA is found in Africa, indicating that this is the centre of origin of the genus *Homo*. Outside Africa there is only one basic global type, also commonly found within the continent. This basic type falls into five related categories approximatly equivalent to Coon's five races of human. The exciting detail of this discovery is that estimates of the time it would take for such changes to accumulate suggest strongly that our common African ancestor existed as recently as a quarter of a million years ago. We may all have descended from an African 'Eve' who lived less than 300 000 years ago. (R. Cann, M. Stoneking and A. Wilson, 1985.)

10.8   The social concept of race

The social concept of race is different in kind from any genetical adaptation view or evolutionary view of human diversity. A group in society with a common ethnic origin, nationality, culture, language or religion may consider itself a 'race'. The term **ethnicity** is sometimes used to describe race in this more social sense. Ethnic groups are generally composed of individuals with a long shared history. They will probably have a common geographical centre of origin, with a common and distinctive culture. Into this group, in the past, individuals may have come by marriage or assimilation. Whereas nationality may be imposed on you by your place of birth or country of residence; your race is much more a matter of who you perceive yourself to be and who you are perceived by others to be. Race, socially speaking, is therefore more determined by perceptions than genetics.

Any citizen of the United Kingdom has a multi-ethnic origin. Periods of migration have in turn brought Celts, Romans, Saxons, Danes, Normans, Jews, Hugenots, Poles, West Indians, Pakistanis, Chinese, Vietnamese, East African Asians and Cypriots. There is no such thing as a 'pure race'. The only thing that therefore defines an ethnic group is a continued sense of allegiance within that group. At its weakest this rapidly results in cultural assimilation and intermarriage; at its strongest the ethnic group may well continue for a long period in relative cultural and biological isolation. The Welsh living in England are a category of the first kind. They may observe St David's Day and be partisan at a Welsh Rugby International but unless they keep up their Welsh language and culture their Welsh identity is unlikely to be long lived.

Where the physical appearance of individuals distinguish races, self-perceptions and group perceptions may last much longer. Because skin colour is such an obvious difference it has taken on a significance out of all proportion to its degree of genetical difference. Although many 'blacks' are more 'white' than 'black' in their genetic constitution, they may perceive themselves as 'black', because of society's attitudes. The history of racial groupings in a society depends to a very large extent upon such perceptions. In the case of 'white'

European peoples that have been assimilated into Britain, many have rejected their minority status to join the majority. For example, some Italians disassociated themselves from other Italians to escape from presumed links with the Mafia. After World War I some Germans anglicised their names to escape from anti-German prejudice. The long term ethnic groups that have persisted have done so by the individuals continuing to define their own identity within the group to which they feel they belong. The 'group' may be defined by culture, language, physical appearances or religion. This feeling of belonging or 'felt membership' is both at the rationally argued and at the emotional level.

That 'race' in British Law is not just a genetic matter is important to grasp. A now famous case was taken to the House of Lords in 1983 when it was ruled that a Sikh schoolboy, Gurinder Singh, could not be expected to comply with his school's rules regarding dress without experiencing racial discimination. The orthodox Sikh religion requires that boys should not cut their hair and that they should wear a turban. Both of these had been objected to by the school. It was ruled by the Law Lord, Lord Fraser of Tullybelton, that although the Sikhs were not genetically distinct from other peoples in the same region of the Punjab, because of their shared history, religion and culture the Sikhs are by British Law a 'race' and thus protected by the Race Relations Act of 1976. The noble Lord also expressed the view that it must therefore be possible to be a member of a 'racial group' not only by birth but by adherence to a creed or by adoption into the group.

10.9 Racial prejudice

Racial problems seem to arise where one group in defining itself as the 'in-group' requires an 'out-group' against which to define its own distinctive identity. This indicates that racial awareness, orientation and attitude is in large part culturally developed and relates to the way we are brought up as children. If the cultural attitudes of a society influence the way members of that society grow to regard their own and other racial groups, then a group experiencing poverty, status deprivation or competition for jobs is more likely to express prejudices (J. L. Watson, 1976). All young people to some extent discover their personality by joining groups. In-group and out-group distinctions are commonly made by young people regardless of whether race is involved; security is to be found in a social group. If, however, the self-image developed within the in-group does so with feelings of guilt, fear or anxiety then prejudice may develop.

Mary Goodman (1966) carried out studies of race awareness in young children by the use of dolls of different colours and with different facial features. Goodman discovered that a **racial awareness** of black and white dolls had developed in more than half the children in the U.K. before they were four. Between the ages of three and six a **racial orientation** developed whereby children were able to give themselves a group identity. In a racialist environment race related words are learned and likes and dislikes at this age are openly expressed. **Racial attitude** is adopted before most children are eight years old. Children will at this age adopt the surrounding adult perceptions of status, ability, character, occupation, wealth, etc. **Racial prejudice** is intensified in the early teens especially where the in-group and out-group identities of teenagers are being established in a social framework of economic frustration, suspicion, distrust, or threatened social status.

Racial discrimination is not the same thing as racial prejudice. People may discriminate unlawfully even though they themselves are unprejudiced. Such direct discrimination might occur where an employer, who is unprejudiced, turns down a black job applicant on the grounds that other employees or customers would not like to have a black colleague. Indirect discrimination is to make requirements that have an adverse impact on an ethnic group. An example here would be that where a Sikh schoolboy might be required to wear a school cap. This would go against his religious requirements to wear a turban.

Many studies indicate that although the media, books, comics, films, etc. have a powerful influence, it is parental attitudes that count most strongly. Where racialism is firmly rooted in the ruling adult culture it is difficult to escape from its long term damaging influences.

Fig. 10.7   Children become more aware of the physical differences between people, such as sex and colour, as they become more perceptive of their environment and themselves. Their subsequent orientation and attitude depends on early experiences in their social environment. These four year old girls at a playgroup are at such a formative age.

11 Eugenics and euthanasia

11.1 Introduction

One cannot escape from the reality that the more we know about the science of genetics, the more we are able to interfere with human life by directing our breeding and reproduction to particular ends. Science by its method is continuously revealing new knowledge. This knowledge very often leads to applications in new technology. What does the future hold when the present biomedical technology has altered out of all imagining? On the one hand the opportunities for relieving human suffering are fairly clear. But may not the control we have at the present be lost, as 'faceless scientists' transform what is the science fiction of today into the science facts of tomorrow? After all, one might argue that the breeding of plants and animals to suit human needs has been so spectacularly successful that we should now turn our attention to our own species and perhaps achieve even better results? If we do interfere, what is it 'right', or ethical, to do and what actions would be 'wrong', or unethical? Faced with the bewildering choices, how are we to choose? Certainly having to make choices is not new. Human beings have always had control over many dimensions of their lives, and societies have for thousands of years developed moral codes to help with controlling human social behaviour. What is new is the pace of change. Let us examine a few questions by way of introduction.

Is it right to induce the abortion of a foetus which is known to be carrying an at present incurable genetic disease? For what severity of genetic disease is such an induction justified? Who has the right to say; the parents, the doctors, the mother alone, or does the foetus have a 'right to life'? If one 'allows' such abortions as socially acceptable, should parents have the right to choose the sex of their child? Why should they not also choose the colour of its eyes, or good looks or intelligence? If infertile couples may have their infertility 'cured' by egg or sperm donation, as already happens, and if sperms and embryos may be 'banked' in a deep freeze, as is also possible, what is wrong with using methods to selectively breed for a 'better' kind of human being? Yet again, embryos may be cloned and then kept deep frozen. Why should we not keep a spare genetic copy of each individual in the bank to provide a replacement set of organs, should our own wear out or become diseased? Such questions may appear either frivolous or selfish, but society may require answers before the possibilities become practices. One may be so revolted by the new possibilities that more genuine benefits that are possible may be rejected. If for example a gene implant to cure muscular dystrophy were possible, would it be right to use an early human embryo to research the implantation method of curing what is otherwise a lethal and distressing condition for sufferers and their families?

This chapter examines eugenics and euthanasia, two important social movements, that have a bearing on such questions. **Eugenics** is concerned with human improvement by genetic means, such as breeding, and therefore relates indirectly to genetic counselling. **Euthanasia** is the act of painlessly and directly terminating human life, and may be considered to include foetal abortion. Both eugenics and euthanasia have a basis of good intentions, but both are fraught with misconceptions and moral problems. The subject of **ethics**, the science of morals, is briefly considered, and some guidelines are suggested for facing the implications of the new genetic technology.

11.2   The origin of the eugenic movement

The earliest ideas of improving human society by breeding go back to Plato's *Republic* (427 B.C.) in which a system of breeding festivals for favoured men and women and state nurseries for the favoured children were envisaged; in Aristotle's *Politics* (384 B.C.) freedom to choose an abortion was advocated, along with the humane destruction of the handicapped and crippled. Two thousand years of Christian culture in Europe have given us our traditional framework of attitudes to such things as marriage. In the last century an awareness of the mechanism of animal and plant stock improvement followed Darwin's evolutionary theory. Society suddenly found itself looking at the human breeding system more critically.

Francis Galton (1822–1911) is the father of the term 'eugenics', which he coined in 1883. His most famous book, *Hereditary Genius* (1869) contains the results of his studies of the family pedigrees of eminent men. He advocated 'judicious marriages' between 'men of genius and women of wealth'. Galton himself had no mean pedigree, being related to a host of eminent Victorians, and married to a wealthy woman. He was indeed a genius ahead of his time but he held ideas about female intellect, social class and the superiority of his race that would be derided as prejudice today. He certainly underestimated the effects of a favourable environment though it is to his credit, in this respect, that he perceived the value of studying identical twins. Before his death he had endowed a research post in eugenics at University College, London and founded the influential 'Eugenics Society'. The most brilliant professor to hold this university post was the mathematical geneticist Karl Pearson, who must take most of the responsibility for the extreme hereditarian views (see Chapter 9.5) that spread to much of Europe and the U.S.A. in the first half of this century. This is indeed a sorry tale.

Eugenics became openly practised and advocated in the U.S.A. in the 1920s and 1930s, though its implementation can have had little long term genetic effect. L. M. Terman, H. H. Goddard and other eugenicists, by insinuating that I.Q. had a largely genetic base, soon had everybody thinking that 'feeblemindedness' and 'criminality' were genetically inherited in family pedigrees. Goddard's prejudiced falsification of data on certain immigrant families was blameworthy. By 1931 over half the States had made eugenic sexual sterilisation Laws, against the insane, the mentally retarded (for whom Goddard coined the word 'moron'), the epileptics and the sexual deviants. Over a thousand people were sterilised, under the most severe legislation, in the State of California. At the same time the belief in the superiority of the 'Nordic' race became more prevalent, leading to quite erroneous genetical ideas of 'racial purity'. As a result, harsh restrictions were placed on immigration from the 'inferior stock' in Italy and Eastern Europe.

Racist assumptions were also widespread in Britain and Europe in the 1930s and coincided with the growth of fascism. The eugenics movement was attacked both by geneticists, such as H. J. Muller (1932) and by literary figures, such as G. K. Chesterton and Aldous Huxley (*Brave New World*, 1932). However, the protests came too late, for Hitler's pursuit of breeding a 'master race' was to lead to the Jewish Holocaust. In 1935 the Nazis deprived German Jews of their citizenship and forbade marriages outside the Jewish community to protect 'German blood'. In the next decade persecution, internment, torture, execution and mass extermination accounted for 6 million deaths. Until the birth of modern genetic counselling and foetal diagnosis, all talk of eugenics was rightly muted.

11.3   Euthanasia

Euthanasia is generally defined as the act or practice of painlessly putting to death a person suffering from a painful and incurable disease. The term euthanasia is derived from the Greek for an 'easy death'; it is commonly referred to as 'mercy killing', but there are many kinds of euthanasia which should be distinguished.

Euthanasia, first of all, may be **active** or **passive**. If active, a step would be taken to end the life, as in the taking or giving of an overdose of drugs to hasten death; if passive, a life supporting action, such as the maintenance of artificial respiration or giving a newborn baby warmth and food, would be refrained from. **Voluntary euthanasia** is where the death results from a conscious decision of the individual to end their life, or have it terminated for them under defined circumstances. **Non-voluntary euthanasia** is where such permission is not given, but the family or doctor makes a decision in what might be considered the best interests of the individual.

In Judaeo-Christian (Western) culture euthanasia has always been prohibited, being against the sixth commandment – 'Thou shalt not kill'. It is illegal in most nations' Law, although successful attempts to put some form of voluntary euthanasia on the statute books of several European countries have begun. The first attempt in Britain at legalisation was in 1936, but it is still not lawful. Euthanasia has perhaps been associated in people's minds with the 'eugenic' extermination of the Jews. It is rightly regarded with suspicion, but are there any grounds for exceptions?

A doctor's duty to a patient is to 'do good, according to their ability and judgement, and not to cause harm' (The Hippocratic Oath). If there is extreme suffering, many doctors feel that this does allow them to 'not unnecessarily prolong life'. Few doctors welcome becoming involved in decisions to end a patient's life, even where pain and helplessness are involved for the terminally ill. In Law, where euthanasia has been proven to have taken place, the death is certified as 'suicide', if self-inflicted, or as 'murder', if performed by another, even on the written instruction of the deceased. The capacity of modern medicine greatly to prolong life has raised the question of whether, in some circumstances, there is not some entitlement to voluntary euthanasia. Some families caring for the elderly are under strain, but is this grounds for legislation? In such family cases the decisions required of doctors would be peculiarly difficult to make. Amongst doctors, passive euthanasia is commonly felt to be 'right' only in defined circumstances, such as the case of *severe* genetic malformation at birth, or where 'brain life' and consciousness are no longer in evidence. From a legal standpoint, if a doctor decides to withhold some life-extending treatment, he or she may be sued for negligence by a family expecting continued life-support.

Most doctors oppose any legalisation of euthanasia because it would undermine their position of trust with their patients; it is probable that few patients would request voluntary euthanasia now that pain-killing drugs are so effective, and those that did might well be in an impulsive or depressive state of mind. Legalisation could put pressure on those suffering to relieve their carers of responsibility; it also might well open the way to all kinds of foul play, state imposed legislation and a weakening of social morality.

Conversely those favouring legalisation of some forms of voluntary euthanasia, such as the 'Exit' campaign, see permitting euthanasia as conferring dignity to suffering people, as being an open moral choice for a rational being and allowing legality for what is at present practised covertly. Many eminent philosophers and religious thinkers feel that under some circumstances it is morally right. Although the subject of euthanasia is most discussed in relation to adults, in our society pertinent moral issues surround human life before birth where severe genetic disability of newborn babies is concerned. Should such babies be given every encouragement to live, or should they be allowed to die for their own peace and the good of their families? Who are we to judge? How are we to decide on such an ethical issue?

11.4   Ethics and choice

Ethics is the science of morals; it concerns itself with human character and conduct. By 'character' is meant such things as a person's ability to see and understand what is right and wrong; by 'conduct' is meant a person's behaviour that expresses whether they have these perceptions. The **morals** of a particular society reflect what that society currently perceives to be right or wrong. Morals are derived from a society's secular and religious teachings and are often reflected in the society's laws. Moral codes certainly change in the course of history, but societies without moral codes of behaviour are likely to break down.

When deciding what is 'right' and 'wrong' there may well be a conflict between the freedom of individuals to seek the best and greatest good for themselves, and the greater good of the society as a whole. The 'autonomy' of a person, that is their individual freedom to act, must be partly subject to the best outcome for the society as a whole. Young children have no autonomy and parents assume the right to make moral decisions on their behalf. As children grow up they take on more autonomy to decide moral matters for themselves. Although we may question the morality of the culture in which we are reared, we tend in the end to adopt the morals around us that prevail. Moral decisions are often difficult, and so we discuss them in our community and reflect on them ourselves before making up our minds. It is common for individuals to surrender their autonomy to others if decisions are very difficult. Soldiers may hope to be relieved of the moral conflict that might arise in killing an enemy, by surrendering their autonomy to their commander. Likewise a doctor's patient may follow the doctor's ruling on the grounds that he or she 'knows best'.

Because moral decisions are not easy and are often very burdensome, most societies are happiest if they have Laws. Law-keeping, for example adhering to the speed limit, may sometimes be irksome to us personally, but deep down we know that speed limits are necessary to ensure the safety of pedestrians and other road users. When changes take place in society re-evaluation of morals and then Laws may take place. Rational argument and sensitivity must be brought to the altered perception of the issue. Many Laws have a deep and tested historical or religious base and hence may be proclaimed as 'moral absolutes'. Departure from these may well lead to 'slippery slopes' down which a society may slide.

11.5 Abortion

Abortion is the termination of a pregnancy, resulting in the death of the foetus. Natural abortions, or miscarriages, occur quite commonly, the majority being developmental failures due to some genetic defect. Indeed it is probable that only one quarter of fertilized ova successfully implant to begin a pregnancy, and there is considerable evidence that twinning at conception is common, one of the two embryos dying at an early stage. Natural selection, it seems, is operating to reduce genetic handicap and reduce the number of young born.

Induced abortion has for centuries been a method of birth control. It is legalised in most countries including the United Kingdom (Abortion Act 1967). Rather more than 100 000 official abortions take place annually in England and Wales, 98% being because a child is not wanted for social reasons. Only 2% are for the eugenic reason of avoiding the birth of a child with a diagnosed genetical handicap. About half of the abortions are done under the National Health Service. Abortion is only legal before the 28th week of pregnancy (the vast majority are before 12 weeks), and requires the woman's consent and the agreement of two doctors that it is necessary for the 'health' of the mother. (This expression is undefined but includes her physical, mental and social well-being.) The trauma of late abortion is considerable to the woman concerned, and often also to medical staff. There are many well-documented cases of babies born under the 28 weeks limit surviving this 12-week-plus prematurity. That survival of a foetus so young is possible is one of the pressures for reform of the Abortion Law.

The question of when a human life begins and the question of when, if ever, it is justified to take foetal life are important ones. In the conservative view held by the Roman Catholic Church and probably by many other people, human life begins at the moment of fertilisation. This is certainly a valid genetic concept, for it marks the genetic uniqueness of each individual's life. (The term 'conception' has no clear biological meaning, being freely used for both fertilisation and implantation.) The logic of the conservative view, however, needs to be better reconciled with the known wastage of embryos at this stage. If fertilisation confers on this tiny mass of cells a 'human being' status and 'soul', then many human deaths are going unremarked and monozygotic twins must subsequently share one soul between the two! Contraceptives such as the 'morning-after' pill and intra-uterine device (IUD) are clearly directed to destroying such early life, although in Law they are regarded as contraceptives and not abortificants (ruling by Minister of Health, 1981).

In the strict conservative view, all contraception is an interference with the start of human life and the zygote, embryo and foetus have a moral equivalence, in 'worth', to a human person. One may disagree with the conservative view, but it provides a moral stand on the slippery slope to which many people hold.

In the most 'liberated' view human beings do not achieve the status of a 'person' until they are self-conscious, rational and can decide matters for themselves. This view would not give any rights to a foetus and would not distinguish the act of abortion from a mother killing her newborn baby. Such a view might deny 'a right to life' to the severely mentally handicapped and brain-damaged individual.

Moderate views lie, uncomfortably, between these extremes. What makes moral decisions difficult is that no sharp lines can easily be drawn. It helps perhaps to have a gradualist approach whereby the embryo and early foetus are seen as a *living human organism*, the later human foetus and newborn infant as additionally a *human being*, and the child emerging from infancy as additionally a *human person*. Such a philosophical view (Lockwood, 1985) needs to be reconciled with the religious insights that see something unique about the human species in the animal world. The gradualist approach does not answer the question 'when does human life begin?' but usefully rephrases it. Many people are revolted by the late termination of foetal life allowed in the Abortion Law and would at least seek to make the latest legal date earlier in the pregnancy; others feel that a mother has sole rights over the foetus that she carries until its birth, and that all such decisions about 'her body' rest with her alone.

11.6 The reproduction revolution

Although some couples may have no children through choice it is tragic that one couple in ten may find themselves to be infertile. Many fail to conceive because of, on the male side, low sperm number, viability or capacity to fertilise, and on the female side, poor ovulation, blocked oviducts or occasionally a poor environment for fertilisation or implantation. Much research has been done on infertility in the past two decades, especially now that the number of children in need of adoption has been reduced by the greater availability of abortion. Because of the development of new fertility techniques an Inquiry was instituted in the U.K. in 1982, which Mary Warnock chaired. The Warnock Report on human fertilisation and embryology was published in 1984, but to date no legislation has resulted, except on surrogacy (July 1985).

In some cases of male infertility, semen from the natural father may be so treated as to make **artificial insemination by the husband (AIH)** a solution to the fertility problem. Where the husband is completely infertile it is possible for an anonymous donor of roughly matched physical features to provide semen. **Artificial insemination by donor (AID)** is quite widely practised, leading to more than 1000 conceptions annually in the U.K. The practice is not illegal but the children are at present regarded in Law as illegitimate. Although it is some-times felt that AID is an intrusion into the marriage partnership and the subsequent child may be upset by the knowledge of his or her origin, a considerable number of couples seek it, and the husbands frequently regard the situation much as an adoptive father might be expected to do.

Female infertility is most often due to ovulation difficulty, blocked oviducts, or a hostile environment for fertilisation. Much research has led to a technique, developed by Steptoe and Edwards, of **in vitro fertilisation** (*in vitro* literally means in a glass vessel, hence the popular notion of the **test-tube baby**). IVF involves the use of drugs that imitate the hormones that naturally stimulate ovulation, inducing the super-ovulation of up to twelve ova at one go. These are removed from the woman's body by a minor operation, employing a fibre optic 'laparoscope'. After a short time, to allow the ova to mature, they are fertilised, most commonly with the husband's semen, before being grown on for two or three days to become a cluster of embryonic cells. At this point a number of healthy embryos are introduced to the uterus where, if the technique is successful, implantation may take place. The number of embryos introduced at this stage is always high, for each only has about a one in five chance of implantation. IVF is an appropriate treatment for about 30% of cases of female infertility. The success rate in starting pregnancies following IVF, and the success in achieving live births from IVF implantation, are each as high as 40%. Thus the majority of women with this sort of infertility problem may success-fully become mothers. The twinning rate is nine times higher than average, because of the unpredictability of the implantation method. Unused embryos may be stored without damage at very low temperatures (cryopreservation) so that subsequent implantation attempts are possible without repeating the IVF procedure.

The first test-tube baby, Louise Brown, was born in July 1978 to previously infertile parents. Zoe, the first child in Britain to be born from a frozen embryo, arrived in March 1985. Both were longed-for children. More than 4000 IVF babies have now been born. Although much personal happiness comes to the thousands of families that are now being helped by these new techniques, IVF has been criticised as being quite unnatural, and poses the ethical problem of what one is to do with the surplus embryos. Our social unpreparedness for these moral problems is illustrated by the events following the death in a plane crash in Australia of both parents of some stored frozen embryos. As a result of campaigning by the Victoria State 'Right to Life Organisation' over 90 women volunteered to give the embryos a womb and a home (*The Times*, 24th October, 1984). This story illustrates the possibility of **surrogate motherhood**. A woman, fertile but unable to have a normal pregnancy following IVF, may engage another woman, as a surrogate mother, to provide a womb for the pregnancy, on the clear understanding that the child born will be surrendered to its genetic mother. Surrogate motherhood may also be used in a case of female infertility to provide a couple with the man's own genetic child, the surrogate mother surrendering the child at birth to the infertile wife. Since the 'Baby Cotton' case (January 1985), commercial surrogate motherhood has been illegal in Britain, under the Surrogacy Arrangements Act.

This revolution in possible ways of reproducing raises many moral questions, not least of which is to ask whether 'rights' are not out of proportion to social responsibility. For many thinking people the technologies now available provide another danger leading to a 'Brave New World' of parents selecting children for a desired phenotype, surrogate mothers rearing children for women evading pregnancy, embryo rearing outside the womb (ectogenesis) and spare part use of cloned embryos. The recommendations of the Warnock Committee and the views of its dissenting members will repay careful study. To cope with the future we shall need both knowledge of the science and a sound ethical sense of the right balance between the greatest good for individuals and that for society as a whole.

11.7   Eugenics and euthanasia in the balance

Where does the balance lie? How can a knowledge of modern biology help us to make up our minds?

The original eugenic concern over the quality of human genetic inheritance arose from the perception that increasingly there would be survival of the less fit; if medical science helps out the weak, surely the genetic strength of the whole population will be the less! Is this view true and how should we respond to it?

Certainly, many of the medical practices we now approve are **dysgenic**, favouring 'less desirable' characteristics. The successful treatment of haemophilia is such an example; at one time male haemophiliacs would invariably have died, but today those that have daughters may afflict their grandsons with the disease. There is better education today amongst those that live with this handicap and we would be arrogant to suggest that they will knowingly pass on the condition. Nevertheless the dysgenic effect is real. Most population geneticists are confident, however, that no discernible weakening of the notional 'genetic stock' will occur in less than several centuries. Indeed, it could be argued that with better control of environmental mutagens things might even get better. High-tech medicine does allow individuals who in the past would have died to reproduce, but such standards at the moment only hold for a small fraction of the world's population. For many in the Third World the Darwinian 'struggle for existence' is ever present.

Eugenic measures are perceived today as of two sorts, a **positive** selection of desirable phenotypes, and a **negative** avoidance of undesirable ones. Let us examine these in turn.

What **negative eugenic** measures, to remove the inheritance or expression of harmful alleles, may we reasonably adopt? The oldest negative measures are the **consanguinity laws** that forbade marriage with close relatives. At least half of your alleles you hold in common with each of your parents and at least half with your brothers and sisters. At least one quarter you hold in common with grandparents and uncles and aunts and at least one eighth with your first cousins. By law you are allowed to marry the latter, but one might remember that from a genetic standpoint it increases the chances appreciably of homozygosity for a harmful recessive allele amongst your children.

Screening programmes for detecting the carriers of particular diseases could be encouraged in the whole population. This is at present done, on a voluntary basis, for Tay Sachs disease, thalassaemia and sickle cell anaemia in certain ethnic groups. In one region of London, where many people are of Indian or Cypriot extraction the incidence of thalassaemia, at birth, was reduced by 78% from 1975 to 1981 by prenatal diagnosis and selective abortion. Gene probes are already on the market for detecting cystic fibrosis and Huntington's chorea in the carrier state. We can expect many more disease carrier states to be detectable soon by such means. Many people may feel affronted by the possibility of such screening and may not wish to have their lives interfered with. Certainly this would be so if people in any way felt 'banned' from reproducing!

Pre-natal screening, by chemical analysis of amniotic fluid, can reveal certain afflictions early in foetal development. Chorion biopsy allows a ready check on the chromosomes that are healthy. Both of these allow decision time for the option of an abortion of a foetus otherwise destined for a life with a predictable handicap. At present most foetuses are not invasively investigated, but the non-invasive use of ultrasound scanning has already reduced the incidence of babies born with spina bifida in the United Kingdom.

It is quite possible that our society may develop a concept of 'genetic health' that could lead to some avoidance of the distress that the most harmful genetic conditions produce. However, knowing that 1 person in 25 is a cystic fibrosis carrier is one thing, wanting to know whether it is *you* is quite another. If you had that knowledge, and could therefore be fairly sure whether you would have a CF child or not, would you be happier? (See Appendix to Chapter 7.) It is quite possible that in ten years time, couples intending to have children will seek voluntary screening for the major genetic handicaps. Where both parents are heterozygous, at one potential disease locus, or the woman is found to possess a disease-carrying X chromosome, they would then, together, be forewarned of the risks. If they decided to have children and were not morally opposed in principle to the abortion of an early embryo, the opportunity of chorionic villous sampling (CVS), early in pregnancy, would reveal whether or not the early embryo would express the defect. It has been suggested that in such cases genetic screening of the pre-embryonic stage, as part of an IVF conception, would allow a normal pregnancy without fear of a handicapped child or the threat of abortion. Such a screening programme would have the potential for good, but many people would argue, with legitimate concern, that we would thereby be interfering with creation too much, and that that would be bad! Before such a change could be decided upon we would need a better morally and scientifically educated population. (We should not forget that the immunisation of children was feared by parents until they saw that the risks were less than the threats of the diseases.)

Positive eugenics, the promoting of the desirable, seems at first sight a questionable practice. Although Plato's wild idea of mating festivals for the favoured youth of Athens may seem far-fetched, **arranged marriages** and all manner of other social customs may be considered to derive from a society wanting the best outcome for its children. Although arranged marriages are often to do with property and status there is no doubt that 'good breeding', in the social sense, is perceived as having a genetic element to it. To some extent this must be a society practising the breeding of its offspring with a positive eugenic motive. (See also 10.3 on sexual selection.) If we are allowed to select our mates freely, should we be allowed to select the sex of our children? The advent of AID and IVF technologies raises the possibility of positive eugenics on the grand scale. Although it has been seriously suggested that Nobel Prize-winners should donate semen to a sperm bank, this would be unlikely to make much impact on people's natural preferences and seems dubious when the qualities they have as people are not provenly genetic in origin! In the proposed legislation arising from the Warnock report it is likely that no AID sperm donor or IVF donor will be allowed to 'give' more than ten offspring by this method. As far as is possible donors would be screened for genetic health and broadly classified by physical appearance. Perhaps the children genetically descended from these anonymous donor parents should have the right to know who their genetic parents are, as is currently the case for adopted children. Since the discovery, by Alec Jeffreys of Leicester University, of the technique of **genetic fingerprinting**, it is possible to compare the probed blotted patterns of different individual's DNA and establish, merely on the basis of a small blood sample, whether two people are closely related or not *and* the degree of closeness of the genetic relationship between them. In this light parents will need to be open about their children's genetic origins, for they will soon be open to test.

Although it is claimed that **embryo research** on 'spare' IVF embryos may assist in ways of improving human health, there are medical scientists and many others who have ethical misgivings about it. The Warnock Report recommends that such experimentation should only be allowed on embryos within fourteen days of fertilisation. Such research could foreseeably result in gaining further knowledge of congenital malformations (deformities that arise in development) and in detecting gene and chromosome abnormalities before implantation.

Genetic engineering is involved in the making of genetic probes and in many of the new diagnostic techniques. One possibility that exists is the making of a **genetic implant** of a missing gene into the genome at an early stage of development. This would involve recombinant DNA technology, producing a coded sequence that could be inserted at some point into the genome of the cell. This has been achieved with the gene for growth hormone in mice, but the genetically 'improved' dwarf mouse grew up into a very much larger 'mighty mouse', three times larger than normal size! Despite such early research problems there is a real possibility of using such **gene therapy** to help with diseases such as diabetes, where a body organ, in this case the pancreas, has failed to produce insulin from its cells.

11.8 Freedom, responsibility and the selfish gene
In concluding a book on human heredity one should finally address the

question of whether the human genetic program in any way restricts our freedom as individuals to act in a particular way. Are we in any way behaviourally controlled by our genes? Is our DNA selfishly driving us to act in such a way as to ensure its own survival? (Dawkins, 1976).)

Much of philosophy and theology argues that a human being has an animal side to his or her nature but that each person can to some extent come through this level of existence to a much higher plane of being. Here, standing outside oneself, one can be more objective about one's own behaviour and perceive its effects on other people. We are self-conscious autonomous beings. We can understand right and wrong and make judgements of a moral kind.

According to the classical behaviourist school of sociology, *all* our social and cultural behaviours are a product of our upbringing; genetic blueprints have little or nothing to do with behaviour. We are born with a clean slate and everything we are as personalities is added by our experience of life. The sociobiology school, which sees much social behaviour as adaptation to confer a Darwinian 'fitness' in the evolutionary struggle, is opposed to this view. With respect to animals, it is convincing that social behaviour has evolved to operate selfishly in favour of the possessor's genes (E. O. Wilson, 1975; Reiss and Sants, 1987). Should what is true for other animals be any less true of ourselves? **Sociobiology**, which in effect implies much less general freedom of human social behaviour than most of us would care to acknowledge, is controversial in the social sciences. Although it is stifling to feel a victim of a 'genetic determinism', it would be fair to say that most biologists hold that social scientists have not yet come to terms adequately with the genetic realities of the human evolutionary past.

It is fair for scientists to do battle over scientific theories, but scientists should not tangle their theories with morality. This does not mean that they are free from moral responsibility. Every scientist is a human being also and human beings have immense freedom to do good! Genuine altruism extends a great deal further than ensuring the survival of your own DNA!

Bibliography and references

Alberts, Bruce, et al., (1983) *The Molecular Biology of the Cell*, New York, Garland Publishing, Inc.

Ayala, F. J. and Kiger, J. A., (1984) *Modern Genetics*, Benjamin/Cummings Publishing Company, Inc.

Burnet, Lynn, (1986) *Essential Genetics*, Cambridge University Press

Cavalli-Sforza, L. L. and Bodmer, W. F., (1971) *The genetics of human population*, San Francisco, W. H. Freeman and Company

Cavalli-Sforza, L. L. and Bodmer, W. F., (1976) *Genetics, Evolution and Man*, San Francisco, W. H. Freeman and Company

Cavalli-Sforza, L. L., (1977) *The Elements of Human Genetics*, (2nd ed.) W. A. Benjamin Inc.

CIBA Foundation, (1985) *Abortion: medical progress and social implications*, Pitman

Clarke, Sir Cyril A., (1987) *Human Genetics and Medicine*, (3rd ed.) Edward Arnold

Connor, J. M. & Ferguson-Smith, M. A., (1984) *Essential Medical Genetics*, Blackwell Scientific Publications

Coon, C. S., (1962) *The Origin of Races*, New York, Knopf

Davies, Kay E., (1986) *Human Genetic Disease: A Practical Approach*, Oxford, IRL Press Ltd.

Dawkins, R., (1976) *The Selfish Gene*, Oxford, Oxford University Press
        (1986) *The Blind Watchmaker*, Oxford, Oxford University Press

Dobzhansky, T., (1962) *Mankind Evolving*, Yale

Edwards, J. H., (1978) *Human Genetics*, Chapman and Hall

Emery, A. E. H., (1983) *Elements of medical genetics* (6th ed.) Churchill Livingstone

Emery, A. E. H., (1984) *An Introduction to Recombinant DNA*, John Wiley and Sons Ltd.

Eysenck, H. J., (1973) *The Inequality of Man*, London, Temple Smith

Feinburg, G., (1983) *The colours of people*, pub. The Commission for Racial Equality, ISBN 0 907920 19 5

Galton, Sir Francis, (1969 re-ed.) *Hereditary Genius*, London, Macmillan

Garn, S. M., (1971) *Human Races*, 3rd edn., Thomas

Gould, Stephen Jay, (1980) *Ever since Darwin – Reflections in Natural History*, Ch. 29, 'Why we should not name races.' Penguin
        (1981) *The Mismeasure of Man*, Penguin (Pelican)

Harrison, G. A., et al., (1976) *Human Biology*, Oxford, Oxford University Press

Harth, D. L., (1983) *Human Genetics*, New York, Harper and Rowe

Humphrey, D. and Wicket, A. (1986) *The Right to die; Understanding Euthanasia*, The Bodley Head

Huxley, A., (1932) *Brave New World*, Chatto and Windus

Ingle, M. R., (1986) *Genetic Mechanisms*, Oxford, Blackwell

Jenkins, John B., (1983) *Human Genetics*, The Benjamin/Cummings Publishing Company, Inc.

BIBLIOGRAPHY AND REFERENCES

Jensen, A. R., (1969) *How much can we boost IQ and Scholastic Achievement?* Harvard Educational Review, 39:273

Kuhse, H. and Singer, P., (1985) *Should the baby live? The problem of handicapped infants,* Oxford University Press

Levitan, Max, (1977) *Textbook of Human Genetics,* (2nd ed.) Oxford University Press

Lockwood, Michael, (1985) *Moral Dilemmas in Modern Medicine,* Oxford University Press

McKusick, V., (1982) *Mendelian Inheritance in Man,* (6th ed.) Baltimore, John Hopkins University Press

Milunsky, A., (1977) *Know your genes,* Penguin Books

Reiss, M. and Sants, H., (1987) *Behaviour and Social Organisation,* Cambridge University Press

Rose, Steven and Kamin, L. J. and Lewontin, R. C., (1984) *Not in our Genes: Biology Ideology and Human Nature.* Pelican

Rostand, Jean and Tétry, Andrée, (1965) *An Atlas of Human Genetics,* Hutchinson & Co.

Singer, P. & Wells, D., (1984) *The Reproduction Revolution: New ways of making babies,* Oxford University Press

Stewart, J. Bird., (1987) *Genetics in relation to Biology,* School Science Review, Vol. 68, no. 245, June 1987

Suzuki, D. T., Griffiths, J. H., Miller, J. H. & Lewontin, R. C., (1986) *An Introduction to Genetic Analysis,* New York, Freeman and Co.

von Cranach, M., Foppa, K., Lepenies, W., and Ploog, D. (1979) *Human Ethology,* Cambridge University Press

Warnock, Mary, (1985) *A Question of Life: The Warnock Report,* Basil Blackwell

Watson, James D., (1968) *The Double Helix,* Weidenfeld and Nicholson
(1984) *The Molecular Biology of the Gene* (4th ed.) Benjamin/Cummings Publishing Company Inc.

Watson, James L., (1977) *Between two cultures, migrants and minorities in Britain,* Oxford, Basil Blackwell

Wilson, E. O., (1975) *Sociobiology: The new synthesis,* Cambridge Massachussets, Harvard University Press

Index

129